AF413766

MAKING SENSE OF LIFE

MAKING SENSE OF LIFE

Develop Your Own Theory for Happiness and Achievement

SIMIN CAI, PhD

with MIKAELA ASHCROFT

Published by Forbes Books, Charleston, South Carolina.
An imprint of Advantage Media Group.

Forbes Books is a registered trademark, and the Forbes Books colophon is a trademark of Forbes Media, LLC.

Printed in the United States of America.

10 9 8 7 6 5 4 3 2 1

ISBN: 979-8-88750-558-9 (Hardcover)
ISBN: 979-8-88750-559-6 (eBook)
ISBN: 979-8-88750-560-2 (Audiobook)

Library of Congress Control Number: 2026901320

Cover design by Matthew Morse.
Layout design by Ruthie Wood.

Since 1917, Forbes has remained steadfast in its mission to serve as the defining voice of entrepreneurial capitalism. Forbes Books, launched in 2016 through a partnership with Advantage Media, furthers that aim by helping business and thought leaders bring their stories, passion, and knowledge to the forefront in custom books. Opinions expressed by Forbes Books authors are their own. To be considered for publication, please visit **books.Forbes.com**.

03-02-2026 1:36

This book is dedicated to everyone.

Of course, to my son and my mother—but also to everyone I have met and everyone I have yet to meet. I respect you for who you are, believe in your potential to achieve more, and have been amazed—or hope to be amazed, again and again—by your uniquely individual stories, written and still in the making.

This book is also dedicated to my late father, and to all those who have inspired me but already left this world, whether or not I had the honor of knowing you personally. I will forever be in your debt for the insights you shared, the legacy you left, and the inspiration found in the lives you lived. The only way I know how to repay that debt is to pay it forward—and this book is one way of carrying that forward.

CONTENTS

INTRODUCTION
The Quest to Make Sense

Why do we live the lives we do? Why do some of us search for success, while others are satisfied with much less?

Why are so many of us not happy?

Every person I've met has been drawn to a different lifestyle. Some enjoy art, while others embrace science and mathematics. Everyone has a different measure of happiness, and so many are pursuing lifestyles that aren't fulfilling. Many of the people who ultimately led unfulfilling lives couldn't make sense of why they seemed to struggle. Even the people who were most satisfied with life still couldn't seem to make sense of everything.

Humans have been asking the question of *why* for as long as we've existed. Philosophical and religious narratives have sought to provide an understanding of nature and more theological questions. Existentialists believe we create our own meaning in a meaningless universe. Stoics believe that living in a realistic world based on reason leads to fulfillment. Buddhists analyze the nature of suffering and relate it to a path of liberation.

While there are so many philosophies trying to make sense of life, I would like to start at the beginning for my own pursuit.

Many years ago, I pursued physics, completing a BS in applied physics from Shanghai Jiao Tong University in 1988 and an MEng in engineering optics and a PhD in physics from Stevens Institute of Technology in 1993 and 1999, respectively. This was a way for me to understand life at its base level. I knew that, at the heart of every human quest, we all want to make sense of nature. The science that breaks down the order of every living thing into its base elements seemed to be the best place to start to answer some of life's most difficult questions.

Understanding the universe at a base level provided some clarity. I understood to some extent how the physical nature of life worked. Physics made sense. How the universe worked made sense. But when I graduated with a degree in physics, the questions kept coming.

Once I got into the business world, I learned that asking the question of *why* was incredibly applicable to business. I was looking at the world in terms of optics, the specialty I acquired with my education. When the company wasn't doing well, I could look at the metrics to discover why. When a particularly difficult business challenge arose and traditional methods to solve the problem failed, I could break down the individual components of the problem to discover why.

But there were some questions that physics just couldn't answer. They were much more personal.

I felt that the best course of action for me was to start a business, grow a company, and become an entrepreneur. But why me? Why did I feel like that course of action would be the most satisfactory for my life?

Those questions carried a lot of weight for me. The hardships in life weren't always pleasant, but they led to satisfactory conclusions, one way or another, most of the time. Physics is the science of understanding nature, but there were an endless number of questions related to other branches of science as well. I soon began analyzing questions related to humanity. On a broad scale, it's all connected. From physics, we developed different kinds of science, such as chemistry and biology. From there, humanity has branched out to social science, and, of course, you may never neglect science's original root: philosophy. All of us keep asking the same question: *Why?*

Unsurprisingly, I turned to physics to find the answer.

Asking *why* starts us on a journey to self-discovery. It breaks problems down into their fundamentals. It forces us to take a step back to fully make sense of life.

I found that taking an approach that is too broad may negatively impact our success and happiness. We start asking the big questions before we analyze the little ones. Some of the most obvious and grand questions we ask are these: What is the meaning of human existence? What makes a life well lived? And how can we know anything with certainty?

Those questions are intimidating and can often lead to fear of the unknown and, therefore, unhappiness. But what if they were broken down to make them more manageable? Let's analyze some examples to help you better understand what I mean.

Instead of asking about the meaning of life, we can start questioning smaller acts: What activities make me lose track of time? What would I regret doing if my life ended unexpectedly? What parts of my daily life feel most meaningful right now? The broad case of human existence often forces us to get lost in the mix. Often, the simplicity of human existence comes down to personal experience. Trying to find purpose is the most meaningful way to answer the broader question.

Instead of asking what makes a life well lived, we can ask the following questions: When do I feel most alive and engaged? What qualities in my relationships make them fulfilling? When have I pushed beyond my comfort zone in meaningful ways? A single, definitive answer that applies to all humanity and spans multiple cultures does not exist. However, a personal perspective often makes it easier to understand.

Instead of asking how we can know anything about the universe with certainty, we can ask questions that ground us more closely to reality: What's the difference between how I experience something on a daily basis and what it objectively is? What thinking process has led me to the most reliable conclusion? What are the boundaries of what I can know directly versus indirectly? These questions are much more aligned with personal decision-making. Though all the answers may be beyond the limit of our understanding, creating a reliable basis of belief in our personal lives can lead to much greater satisfaction and help us rely more on realizing objective reality.

Once we've answered some of the most fundamental questions, understanding our reality through personal perception, we can tie what we've learned to personal desires. Aligning reality with those desires is what leads us to happiness that we can feel and enjoy.

My personal approach to pursuing happiness in life is based on making sense of life one step at a time. It's a philosophy I've tied deeply to my desire to impact others' lives positively when I have the opportunity. I've found that the more I understand my nature and what I want, the easier it is to make decisions. Stress has been much less of an obstacle in my life. And, by sharing my philosophy with others, I've found that such in-depth connection has led to my own resiliency and effectiveness.

This book is broken down into some simplistic ways of understanding your life. I've included an example that represents different phases of life in every chapter. In the first part, the examples are more closely aligned with the first two decades of life. In the second part, the examples break down common adult experiences. The final part more closely aligns with senior life. It's much like how humanity has grown up over time.

However, even though the narrative follows a typical human trajectory, every moment of discovery is possible at any point in life. In fact, many of us rediscover and reanalyze our situations throughout life. You'll see examples of people who have completely reinvented themselves after discovering something new, even in their later years. You'll also see examples of young people who have found their purpose early and stuck to it for their entire lives. We lead entirely different lives, but we all experience joy, sorrow, and limitation.

Optics is a big part of my life, whether it's related to my education or my career. Where we see an image is not always exactly where it is. Likewise, our understanding of life can be warped by various perspectives, different thinking, and unusual experiences. It's my hope that, when you read this book, you'll take a step back and look at the lens you use. Things aren't always as they appear.

Life can get complicated, and that prevents us from seeing the broader picture. We are living, breathing achievers. No matter where you are in life, give yourself the chance to pursue happiness. Ask yourself the right questions and dig deeply into your own reality. You never know how far this book can take you unless you're willing to give yourself a chance.

PART I:
Making Sense

CHAPTER 1
Making Sense of Nature as Humankind

Despite our vast differences, we all have one thing in common: our desperate need to make sense of life. It's in our very nature to try to understand what's happening to us. From the day that we're born to the day that we die, our need to learn, or curiosity to know why, stays intricately attached to us.

When we get older, we often take for granted the steps needed to understand. In fact, you might not even notice that extra second you take to look around you before stepping out into the street or turning the key to unlock your home. You've done it so many times now that you understand how processes work the first time. It's all second nature.

But when was the last time you thought—I mean *really* thought—about why you do what you do?

If you feel you've heard this all before, you're probably right. It's something that humans do to make scientific discoveries. Slowing down to analyze the world around us is something people have done for centuries. Everything from the discovery of gravity to the understanding of genetics came from the simple question: Why?

The great scientists of the past made incredible scientific breakthroughs using simple steps to make sense of nature, but making sense of your own life doesn't seem to be that simple. Or could it be?

The Discovery of Humanity

When you were first born, the world around you made no sense. Your eyes still needed to adjust, and you couldn't see much past your nose. When your mother came closer, she came into focus, but just barely. Color hadn't fully materialized for you, so any stark changes in contrasting hues made you jump. The feeling of touch was new, revolutionary; it made you want to reach out and try new things. You could just make out the sounds of the people around you, but anything new was frightening.

Can you blame yourself? You were just starting out.

As your vision began to clear over the months, you found the need to try new things. Your blurry vision came into a little more focus. It's also the time when the game of peekaboo became most distressing. Below the age of six months, you weren't quite sure if your grandmother would come back after hiding behind her hands. When she did? A revelation.

Over time, those familiar patterns found their way into your own mannerisms. When your father smiled, you did, too. When your brother laughed, you laughed right along with him. It was finally starting to make sense. You were getting the hang of it. What someone else did, you could do, too.

Your first word came as a revelation to your family, but it only *made sense* that the woman who kept coming back to care for you fit the word *mother*. That fluffy thing in the room that sniffed your face and licked your toes was *dog*. It had only taken a year, and you'd already begun to think symbolically. Everything in the room had a name, and, perhaps, everything outside of the room had a name, too.

Your desire to crawl probably came from your desire to think outside the box. What could you get into? What could you see, hear, feel that was different from what you'd ever seen, heard, felt before? It was the beginning of your quest for knowledge, and something as simple as figuring out how to open a door wouldn't stop you.

In fact, nothing could.

Progression: The Consequence of Human Nature

From an early age, you learned that you could manipulate the things around you to make sense of the world. The way you interacted with nature taught you how the world worked. It was in your nature to progress, and that same desire to progress continues to prevail throughout life. Even if you feel like you don't have the same push to stand as you did when you were young, other areas of your life have drawn your attention, and they still do.

Humanity's desire to discover *why* drew a young Isaac Newton's attention out the window one morning to see an apple fall from a tree. Though he'd seen countless examples of objects pulled down (including his own feet), he felt the urge to ask *why*. Why did the apple fall directly to the ground instead of moving side to side? Why didn't it float upward? Why, when it was on the ground, did it stay on the ground? Surely there was something that kept the apple there.

Now, if that were true, maybe it had something to do with the earth itself. Or maybe it had nothing to do with solid objects and had to do with something else. Maybe there was something invisible that kept the apple on the ground and continued to pull it down even after it had landed.

Newton's obsession with asking *why* led to countless break-throughs in the understanding of gravity, but his innate curiosity is hardly unique. Progress is human nature. Others learned of Newton's findings and incorporated them into their own theories. Improvements in technology led to increased productivity. Improvements in productivity led to more questions. More questions led to newer technology and newer solutions.

All because so many people tried to make sense of nature.

Understanding Prediction and Control

Our development of scientific advancement has come down to one thing: the need to learn more. Ever since the earliest recorded history, there has been a voracious need to understand why. Early Babylonians, stretching back to the seventh century BC, looked up at the skies to find the patterns in celestial rotations. Hippocrates and his school based their work on observation and documentation to lay the groundwork for modern medicine. Euclid's *Elements* came about as the pursuit of knowledge related to ancient ratios.

As the centuries have progressed, the innate desire to learn more, to understand more, has not dwindled. If anything, looking back has given us the foresight to know that there are limitless answers. We keep asking *why*, and there are still things to discover.

Not all the explanations that humans have come up with over the years are up to snuff. Over the years, there have been quite a few ideas, labeled as science, that are merely explanations based on conjecture. For example, in the sixth century BC, Thales of Miletus sought a natural explanation for fluid motion in the natural world. He concluded that water was a fundamental substance in the universe. That answer is not quite right considering what we know now, but it was one of the earliest recorded attempts to explain the natural world.

Both examples of scientific innovation and attempts at rational thought showcase a key principle for understanding the world: Even though there are an infinite number of rationalizations, at some point, you have to stop and settle on one that makes the most sense for all you know so far. There is a point at which we simply don't have enough technology or rationalization to explain what's happening. We have limitations. And, if you keep asking questions, *as an individual*, you may not benefit much from the conclusions considering the effort you may have to put in. You've reached the end of practical or theoretical limits, and questioning more just becomes unproductive.

The development of science and math came as a result of asking these endless questions, from generation to generation, by humankind collectively. The science and math we have written in textbooks today aren't necessarily a conclusion of that work but merely a byproduct of humanity's incessant curiosity.

But the lives of people who lived in ancient times can sometimes seem like a snapshot that isn't realistic today. Of course, that isn't the case. To prove it using something that everyone experiences daily, let's break it down into something a little more manageable.

Imagine you're waiting in traffic to go to work. You've been sitting in your lane, motionless, for the last hour and a half, and you're already late for work. There could be a few reasons why traffic hasn't moved for a while: Maybe there was an accident that is blocking multiple lanes, maybe everyone decided to be on the road that day, or maybe there is a convoy of semis carrying temperature-controlled milk and trying to move as slowly as possible to maintain their cargo.

The point is, as you're sitting there, tapping your thumbs angrily against the steering wheel, you're coming up with a variety of rationalizations for why the traffic has stopped. As you're sitting there in traffic, your mind might go to increasingly bizarre places to rationalize why all the vehicles are at a dead stop. If you've been waiting there long enough, you might start to think that a zombie apocalypse is a reasonable explanation.

But, even if the vehicles don't move for hours, there is a point at which you stop thinking. Maybe it's sunstroke that has made you even more irrational. Experience and knowledge of how the world works tell you that it's unlikely that a horde of the undead is out to get you.

As you stare through your windshield, thinking about how late you'll be for work, you'll likely settle on the most reasonable explanation: There was probably a traffic accident.

Ultimately, we'll choose an explanation that is most consistent with our experiences. These axioms, the conclusions we've drawn from our experiences that eventually become the assumptions that we believe to be true for sure—even without the necessity of proving them to be true—are the foundations for our understanding of nature, the starting points for our reasoning and argumentation. If you've spent a lot of time in traffic outside of rush hour, you'll start to assume traffic accidents are the most likely culprit for slowed traffic, an axiom you've created from experience. If your watch is constantly breaking, you'll assume that it's your watch's fault that you were late for the bus when you didn't check the time on your phone. If the bus is often early because the driver is overly punctual, you'll assume you missed the bus because of the driver's work ethic. If you see people rushing to get on the bus every morning, you'll assume that most people are in a hurry to make it to their locations.

Even if there's no definitive proof that any of these things are true, you'll still come to a conclusion based on the axiom you created from your experiences.

Axioms are premises that you can engage in deductive reasoning. Your reasoning comes from the conclusions you draw. The more you experience and the more your conclusions are proven correct, the more certain you are that the axioms you have created are universally true to your experiences. They act as checkpoints to ensure that what you have determined is truth actually lives up to that definition.

To those who are eternally curious, it may be appealing to keep questioning. If you found the answers to some of life's most curious questions by constantly questioning why, who's to say that you must stop there?

There is only so much that we can experience. Our perspectives are limited, and it's impossible to see things from an infinite number of new perspectives. Despite any attempts to continuously experience new things, there is no point at which you'll experience everything. There is an inherent limitation to life. Endlessly reaching for new possibilities without ever hoping to find a conclusion doesn't help.

Searching for the reason why could give us an added benefit. For example, if we assume traffic is congested because of an accident on our current route, we could find an alternative road or get off at the next exit to avoid the accident altogether. However, if we spend way too much time searching for that *why*, we'll no doubt come up with solutions that have no bearing on what we do. For example, heavy traffic due to Labor Day usually has no bearing on our frustration with traffic. There's no way to change the outcome. Unless you find it amusing to keep searching for the *why*, you'll likely fixate on the most satisfying answer.

Pushing the Boundary

This process of testing axioms is how we have developed science for millennia. We build knowledge by establishing a theory, we predict the outcome, and then we experiment to find if the results are consistent.

It's not in human nature to stop looking. If you look at other animal species, there is a point of contentment, at which there is a balance in social and physical environments that is sufficient to keep them going. Rough conditions mean movement to a new situation, a new environment. But it's humanity that doesn't accept the basis of a no-win scenario. There are always limitations to theories, there are always exceptions to rules, and there are always possibilities that axioms today will no longer be axioms tomorrow because they are only true under some conditions.

But that simple act of progression was not the end. Isaac Newton famously said, "If I have seen further, it is by standing on the shoulders of giants."[1] When creating his own theories, he saw the limitations of previous theorems.

Years before he established his own theory on gravity, he evaluated the great thinkers of planetary motion. Johannes Kepler described the motion of planets around the sun using three distinct rules: (1) planets move in elliptical orbits with the sun at one focus, (2) the line between a planet and the sun sweeps out equal areas in equal times, and (3) the square of a planet's orbital is proportional to the cube of its average distance from the sun.[2] Even if you're unfamiliar with many scientific and mathematical rules when it comes to how planets move around the sun, you can likely appreciate Kepler's use

---

1   Isaac Newton to Robert Hooke, February 5, 1676, in *The Correspondence of Isaac Newton: Volume 2, 1676–1687*, ed. H. W. Turnbull, (Cambridge University Press, 1960), 416–17.

2   *Britannica*, "Kepler's Laws of Planetary Motion," last updated November 27, 2025, https://www.britannica.com/science/Keplers-laws-of-planetary-motion.

of direct observation. There was more than just a coincidence. There was a rule, a law of nature, that ensured objects in the solar system moved in the same patterns over time.

Galileo Galilei, Kepler's contemporary, observed that falling objects hit the ground simultaneously—discounting air resistance, of course. His now-famous experiment of dropping objects off the Tower of Pisa has gone down in history as one of the most influential experiments. He observed a force, albeit one he couldn't identify, that consistently produced the same results.[3] There was something, in essence, that mathematically required both objects, dropped from the same distance, to obey the same laws.

When it came time for his own experiments, Isaac Newton determined that, no matter which object he dropped, it would fall to the ground. The experiment was so predictable that there was only one result every time he dropped something. He expanded his scope to include other elements such as forward motion, predicted the outcome, and tested his theory to see if whatever he dropped would still fall predictably. When he got results, he pushed the boundary even further.

Newton explained in his 1687 published work, *Philosophiae Naturalis Principia Mathematica*, that the development of his gravitational theory wasn't the end.[4] There was something deeper, something more than a single force that pulled downward. As he dove deeper into the mathematics of gravity, he developed three laws of motion that are now so well-known that you probably learned about them in high school: (1) the law of inertia, (2) the law of acceleration, and (3) the law of action and reaction.

---

3   Michael Segre, "Galileo, Viviani and the Tower of Pisa," *Studies in History and Philosophy of Science Part A* 20, no. 4 (1989): 435–51, https://doi.org/10.1016/0039-3681(89)90018-6.

4   Isaac Newton, *Philosophiæ Naturalis Principia Mathematica* (Royal Society, 1687).

It wasn't until the late nineteenth century and the early twentieth century that there were scientists who pushed those boundaries.

The limitations of the human eye and the technology to see things at a much smaller level halted progress until the year 1900, when Max Planck suggested that energy could be broken down into even smaller amounts. Five years later, Albert Einstein suggested that light consisted of small packs of energy he would later call photons. And in 1913, Niels Bohr developed a model for the atom: matter broken down to its basest elements.

But with sizes this small, things behaved a little differently. Isaac Newton's classical way of describing motion didn't pertain to the subatomic. There was, yet again, a flaw in the system. At slow speeds, Isaac Newton's principles worked. But if you get down to the smallest sizes only viewable through a highly intense microscope, those laws seem to go out the window.

The whole purpose of science, as illustrated by this example, comes from expanding scope. Isaac Newton's discovery of the laws of motion was revolutionary. They fit the requirements of the system, which was limited by size and speed. But, when the scope was expanded, those laws were changed to fit the new scope.

Understanding the World Through Humanity's Eyes

Of course, our attempt to understand nature and the universe is at the farthest reaches of our pushing the boundary. At the heart of understanding the universe is understanding ourselves, which is where most people get hung up on the problems that are associated with an ever-moving target. The laws that we have for nature very rarely apply to every person.

René Descartes posed the philosophical phrase "I think, therefore I am." Trying to understand nature and the human condition comes down to assumptions and axioms created from a lifetime of experience. We use our inner thoughts to understand reality. It's through those perceptions that we start to understand the other people around us. This defines our concepts of freedom, responsibility, and the nature of the human condition.

Every era has come up with its own axioms to follow. The philosopher C. S. Lewis said, "A man does not call a line crooked unless he has some idea of a straight line." The basis of social science has come from our perceived sense of right based on the experiences that we and the people around us have had.

The further we dig into any subject, the more we can learn from it. This may benefit us with more perspectives and more subjects to study, as long as we have enough time. As individuals, we have much less time than we do collectively as humankind. However, whether as an individual or as part of humankind, why do we study this subject? There must be a purpose, whether it's to get out of traffic quickly enough to be at work on time or to improve productivity for prosperity. Such purpose brings the set of criteria for us to judge something to be "good" or "bad." But it's not easy. Humanity isn't an easy concept to bottle up. An axiom may be considered good in one scenario and bad in another, simply because the purpose may vary, not even to mention that the axiom may not be an axiom anymore in another scenario. And that is why it is so important to continually search for answers.

Socrates was among the first to promote critical thinking. Identifying the contradictions in someone's arguments and thoughts leads to deeper personal understanding. It was a philosophy on the nature of justice and the idea of moral character. Confucius promoted the idea of cultivating virtues that increased integrity and contributed to the well-being of society. During the medieval period, St. Thomas Aquinas focused on establishing a set of moral principles that came from human nature and reason. Faith and reason were complementary, he said. An understanding of the natural law and the rationale of creatures defined by their surroundings and their faith led to a moral life. Today, philosophers such as John Rawls view the principles of justice through the lens of ignorance of one's own experiences. They set the stage for the political nature of humankind, and the understanding of justice comes from a wide variety of philosophical viewpoints.

All these viewpoints may seem vastly different, but they are a product of their times. Developing an axiom that defines the correct way to live is so subjective and so dependent on everyone's individual experiences that it's difficult to find a unifying solution.

Richard Feynman suggested that understanding nature was a lot like observing chess played by gods.[5] We don't initially know the rules, but if we watch long enough, we eventually figure them out. When we see pawns moving forward in limited spaces, we assume the rest of the pieces do the same. But when we see the bishop moving diagonally, we have a whole new understanding of nature.

It should be a pleasure to continue to push the boundary, to solve how the human conditions that work well for small pockets of society can be extended to all humankind.

Your Corner of the World

Though you probably don't remember the first time you tasted sugar, it likely made an impact on your life. The sweet flavor, paired with other textures and tastes, made sugar one of your favorite foods. When you were a child, sugar probably seemed like the best possible food group. Sugar at breakfast, lunch, and dinner only made sense. The taste of

---

5   Richard P. Feynman et al., *The Feynman Lectures on Physics*, vol. 1 (Addison-Wesley, 1964), sec. 2-1.

broccoli wasn't nearly as delicious, even though your mother kept pushing it on you. You probably thought that, if sugar tasted so good, it must be good for you.

That kind of thinking isn't unusual for children under the age of five. And, when you're that age, even though you don't have many experiences, you still think you know best. Because sugar tasted so good, it *was* good for you.

Even as a child, you created axioms based on your experiences. Your scope was extremely small, so the information you had was limited. You didn't count calories, and you certainly didn't know what macronutrients were. Your limited scope determined that, based on your experiences, you knew exactly which foods made sense going down your throat. To you, tasty meant good, and disgusting meant bad. That was your set of criteria for you to judge good or bad. Therefore, medicine had to be bad, as it tasted so disgusting.

It's much the same at all stages of life. We start with such a limited scope. What we know is confined to what we experience in our small corner of the world. Making sense of nature is confined to how things respond. When you're in your twenties, you may think that putting dish soap instead of laundry detergent into your washing machine will work out the same since they're both soap. When you're in your thirties, you may think refusing to let your kids leave home will keep them from getting sick. When you're in your sixties, you may think that your joints can handle a downhill mountain biking competition.

Wherever you are in life, you're still trying to make sense of nature. Your scope may expand for some areas of your life, but the older you get, the more you're still learning that there's so much you don't know about your nature.

Chapter 1 Questions

- *How have you developed a limit to scope in your own life?*

- *What axioms or beliefs have you developed through your experiences, and how are they related to your reality?*

- *What limitations in your perspectives do you see, and how do you expand your scope to include new ideas once you've found a flaw in your axioms?*

CHAPTER 2

Making Sense by Establishing a "Theory"—a Self-Consistent System

As we flounder through the concept of learning more about life, we start to learn just how limited our understanding is. It would be incredible to know everything we need to do through observation, but it just doesn't work like that.

Luckily, some things just come easy.

We're able to deduce simple conclusions just through observation: When we've lived enough years, we start to notice that dark clouds usually bring rain and sunny skies usually mean the day will be warmer. We can determine animal behavior—for example, when we see a dog wagging its tail, we know it's friendly. We know to stay away from our neighbors when they start coughing uncontrollably or wiping a hand across their noses. Even at a young age, we realized just through observation that smiling indicated happiness.

Though many animals may be aware of the simple observations that come from experience, it's only humanity that truly questions why. We create a grasp of reality to find conclusions to our questions. Even those who have little to no experience in life—anyone from small children to people who just don't get out of the house—are still able to create some logic frameworks, which we may call theories based on experiences.

Perhaps the most unique thing about humanity is its ability to reach a self-actualized state. We're better able to create more standard methods of understanding the world around us by establishing a higher level of psychological development. In essence, you can realize your potential, your capabilities, and your talents to the maximum possible, or to the fullest, as some may express it. You're acutely aware of your own personal growth, and you're always working to maximize your fulfillment.

Everyone throughout history has used this pursuit of growth and the idea of a personal journey to develop theories about nature. How far can someone expand? What limitations can you not overcome? Though we develop a sense of self-actualization when we're young, we keep asking those questions for the rest of our lives.

At some point, we accept something that works for us.

But what if things become slightly more complicated? There are tried-and-true outcomes that we can expect from experience, but what if we've never experienced something before? Or, more curious yet, what if *no one* has?

And only humanity questions what things are like from someone else's perspective. The combination of self-realization and the curiosity to look for new perspectives to develop our own ideas and beliefs is the first step in creating a self-consistent system.

But the question remains: How far can we go?

The Cognitive Climb

By the time you were two years old, you'd started on a new path. No, you weren't up there solving algebraic problems with the best of them, but you were experiencing an incredibly rapid rate of growth in cognitive development and problem-solving.

Think back to this if you can: As you were just breaching two years old, you outgrew the peekaboo phase where you believed the people around you disappeared when they put their hands in front of their faces, but you still felt as if the world disappeared when you covered your eyes. It was a period of time when you had difficulty understanding perspectives outside of your own little world.

Fast-forward a couple of years, and you might have eagerly fed your stuffed animals or G.I. Joes to make sure that they weren't hungry when you left them. You were starving when your mother didn't give you lunch at noon, right? It only made sense that your teddy bear needed to eat, too. As you grew your knowledge of outside perspectives, you started to think that maybe all objects, inanimate or not, had feelings. Of course, it was just a theory of yours at that time.

The older you got, the more you saw things in a new light. It was almost like you were seeing things through your parents' eyes. Maybe you crinkled your nose at broccoli just like your mom did. Maybe you started screaming at the TV when your dad's favorite sports team started losing, just like he did. By the age of five, the adults around you started saying the same thing over and over again: "My, aren't you turning out just like your parents?"

But, of course, that stage didn't last long. As you entered elementary school, you started seeing other mannerisms and other ways that children and adults behaved. Slowly but surely, you started thinking about things a little bit differently. *What do* I *want?* Though you may not have known it at the time, you were developing your own theories about the best ways to live. You were developing a unique personality. At the same time, you were developing other skills as well. Your pattern of thinking involved a complex problem-solving method. The skills you developed didn't just come from one source anymore. The more you experienced in life, and the more ways that you experienced success and failure, the more you started to understand the world differently. You were becoming *you*.

By the age of eight or nine, you started developing ideas associated with a higher level of thinking. Maybe the next time you played checkers with your dad, you blocked the "King me" strategy he'd been using for over ten years. *Strange*, you might have thought, *I've seen him do that a thousand times before, but this was the first time I picked up on the strategy he was using.*

You may have noticed by the time you were twelve years old that things hit a little differently when listening to a story. Back in the time of your earliest recollections, the fairy tale of Hansel and Gretel was mesmerizing for its colorful imagery. But, as a preteen, you started wondering just how realistic it is to trick a witch into getting into the oven.

From the age of two to twelve, your mind was going through something rather extraordinary. Human cognitive development speeds up rapidly. You went from the stage of barely understanding what your eyes took in to developing theories on life. You moved from a stage of obliviousness to one of self-actualization. By the time you reached twelve years old, the way you lived life was based on your response to the theories you developed through your limited experiences. This unique period of your life set the foundation for how quickly you came to conclusions. If things made sense, you just stopped looking.

You probably never again felt the same sense of wonder.

A Human-Centric Vision

And so it was for thousands of years.

The idea of developing a theory based on observations and mimicry has been passed down through the whole of written—and even oral—history. The works of science were deeply influenced by the philosophy and religion of humankind. Indeed, you'll likely have seen the original philosophers equating mathematics and science with philosophy and religion. All these facets of the human experience moved forward together, and the understanding of humanity in the universe moved at the same speed.

When Copernicus postulated that the sun was at the center of our solar system instead of the earth being at the center of the universe, it was met with controversy.[6]

By 1543, the idea of the earth moving along an axis as it rotated around the sun wasn't necessarily new. In fact, nearly two thousand years earlier, philosophers and scientists had postulated roughly the same thing.[7] However, at that time, the observational tools used to make such a statement were rudimentary at best. Other theories better aligned with what others believed at the time. Because humanity was central to religion, an earthbound perspective was the only one that maintained solid evidence.

The writing on a heliocentric model was extremely downplayed in favor of the geocentric model that was most commonly found in

---

6   "Copernicus: Facts, Model & Heliocentric Theory," *HISTORY, https://www.history. com/articles/nicolaus-copernicus.* Copernicus was fearful of sharing his findings and only published them shortly before his death. The Vatican eventually banned *De revolutionibus* in 1616.

7   *Britannica,* "Aristarchus of Samos," by James Evans, *https://www.britannica.com/ biography/Aristarchus-of-Samos.* Aristarchus of Samos was the first person to place the sun at the center of the universe. This theory was forgotten for centuries until Copernicus revived it in 1543.

medieval astronomy.[8] Copernicus's studies while at the University of Krakow sparked his interest in a more innovative way of thinking about astronomical movement. It was enough to get him to focus more on the heavens.

Painstaking hours of recording the positions of stars and planets revealed some inconsistencies and complexities in the typical geocentric universe model. Why, for example, did planets move in motions inconsistent with the rest of the stars? The geometric model was just too simplistic. A model that focused the sun at the center of the solar system instead was much more mathematically straightforward.

Looking back at history, Copernicus's introduction of the heliocentric solar system model was revolutionary, but maybe it wasn't entirely unexpected. The same questions, the same ideas, had merely been buried for centuries. Copernicus unraveled a mystery he had found in the inconsistencies of then-current models. A model that suggested that the earth was at the center of the universe simply didn't jive with the mathematical calculations of people who had been dead for millennia. It wasn't *consistent*.

It simply took the knowledge and fine-tuning of tools across two millennia to provide the way to gather the right data to prove it. And now, we know that the solar system also rotates in the galaxy.

An Informed Decision

With the benefit of hindsight, it might sound fantastical that it took so long for us to believe that the earth wasn't the center of the universe.

---

8   *Britannica*, "Ptolemaic System," by Alexander Raymond Jones, *https://www.britannica.com/science/Ptolemaic-system*. The Ptolemaic geocentric model created by Claudius Ptolemy around 150 BC suggested that the universe orbited a motionless earth.

You might even think that humanity simply gathered data to satisfy its egotism.

But that's a little harsh.

The truth of the matter is that there wasn't enough data to provide solid evidence for one model over the other. True, society and religion backed one theory over another, but there wasn't a widely held understanding of how the earth interacted with the universe. Each scientist developed a theory based on how they interpreted the data they received. The more patterns followed those theories, the more substantiated the theories became.

Let's think about it this way: Imagine you've spent your entire life living in Europe without the benefit of seeing any books or web pages that showed other parts of the world. You've seen horses running through pastures. You've helped raise your brother's bloodhounds since they were pups. For all you know, the animals you've seen every day are the only animals there are.

Now let's imagine that you've sailed to Africa. As you reach the savannah, you catch a glimpse of zebras for the first time. You've seen people ride horses hundreds of times and have even trained some horses yourself. You might think a zebra could help carry some of the cargo you brought. As you head toward the herd, you hear a strange giggling noise to your right. A strange-looking dog with a nasty grin peers at you. *Perhaps*, you think, *it's been in the hot sun for too long. Horses and zebras, bloodhounds and hyenas, what's the difference?*

As you approach the zebras, you quickly learn.

Before anyone can stop you, a rogue zebra breaks from the herd, runs to you, and bites and kicks you until you fall to the ground. As you try to crawl away, you glance at the hyenas now circling you. *Huh,* you think. *Perhaps there are some differences after all.*

Though you might have realized it too late, you've developed a theory that perhaps you can't always predict the outcomes of encounters with creatures that look like horses and four-legged beasts that look like dogs.

Making It Known

Still, despite your newly gained knowledge, neither you nor the people with you can form a coherent conclusion without the use of language. The theories you develop, the axioms you produce, are developed through a complex system that is unique to individual languages.

Gaps in understanding of data and logic ultimately come down to the ability for expression. Throughout history, people who have gathered the right data and come to the correct conclusions have been unable to verbalize them simply because there were no words in their language to describe them.

Though wrestling with ideas not found in current language is frustrating, it's necessary for understanding. Logic doesn't often follow the evolution of language at the same rate. It can take hundreds of years, if not thousands, to develop a system to describe the theory developed from logical pondering.

It's not unsurprising, though.

Let's do another thought experiment. Imagine something existing far beyond the reach of what we can currently understand in space. Think of something outlandish, wild. Imagine how this object would behave in relation to the things around it. Perhaps the object is alive or exhibits motions that represent free will. Now attempt to explain it in a single word. Put a name to the thing that you just created. You can likely explain bits and pieces of your imaginings using current language, but what you have imagined in your mind doesn't yet have an official name. It's not linked with our current language. So, although attaching it to a word seems frustrating, it's the only way that we can understand the data presented to us.

This is why there have been such extensive changes in language over time. The more information we gather, the more we can put down in words. Unfortunately, the variability in language means that not all theories will develop the same way. Clarity of spoken language is clarity of thought.

The axiom developed through constant changes and manipulation in language is its own kind of experimentation. The current axiom is a condensed version of many strings of thought brought together to describe collected data.

Establishing a Coherent Theory

Once a theory is developed and put into words, it must pass through a second test: coherence. In essence, is the theory logically connected to other available data?

Like developing an axiom through language, that theory must be tested often to come to a reasonable conclusion.

This part of establishing a theory involves a lot of trial and error. It's the experimentation phase that lets us know if our theory stands up to criticism. How do you know that a raw egg will splatter on the ground when it's dropped from a building? Well, if you've thrown eggs off buildings a hundred times with the same splattering result every time, you've probably proved yourself right.

If, on the odd chance, your experiment yields ninety-nine splattered eggs but one that is still pristine, can you accurately say that all eggs thrown from buildings will react the same way? If you stopped there, you might assume that your theory was falsifiable or even that your experiment was flawed to begin with. But, if you take a closer look at the pristine egg left intact and you find that, instead of the brittle shell of a chicken egg, it's a granite egg-shaped rock, perhaps your theory is coherent after all.

After realizing that dropping eggs will almost always yield the same results, you can expand your scope and criteria. What would happen, for example, if you could slow the momentum of the egg? Let's say, instead of dropping an egg directly from a building, you attach a tiny parachute that deploys when it's halfway down. Will the egg survive then? Or, if you fashion another experiment that assesses the impact force required to break the egg, how many rolls of toilet paper would it take to surround the egg and protect it when it collides with the ground?

This high level of scrutiny, though tedious if you're trying to find a coherent theory, illustrates our history of hard-fought conclusions. There have always been doubters. There have always been theories proven incorrect. This natural skepticism humanity has for the interpretation of new data is among the most unique and valuable aspects of humanity.

Testing the Theory

The term *judgment* often gets a bad rap. However, it's the only way we can effectively explain circumstances and predict future outcomes.

After we've developed a theory and experimented or tested that theory, we make a judgment call based on what makes the most sense. That judgment is the final say based on the criteria, scope, and testing parameters. In the end, it's what makes the most sense to us.

The judgments we establish are the foundation for creating a beneficial system for our lives. We improve ourselves based on the conclusions we've previously reached. When we develop skills based on those conclusions, we can start tweaking solutions to develop a more consistent life. As we grow, we realize that circumstance A will lead to consequence B. If we can start eliminating outside elements, we can more effectively reach consequence B in each circumstance A.

If we find that the judgments we've made from previous experiences don't lead to beneficial outcomes, we change them. Conclusions are never perfect. We're only human!

As we go through life, we're constantly developing theories and formulating judgments based on the conclusions. It's human nature to constantly try to make sense of life, both collectively and individually. The judgments we generate are vital to helping us better understand ourselves and humanity as a whole. Mistakes are one of our most valuable assets because they lead to better judgments, better conclusions.

A Better Theory

Maybe it's no surprise to you, but creating a more logical structure for your theories is the key to living a more organized life. Plenty of people develop conclusions that have a paper-thin structure, which leads to a life of chaos.

More experiences in life allow you to perfect your system by limiting conclusions to different scopes. You can't always come to the same conclusions based on the circumstances if the scopes are vastly different. It's up to your developing system of problem-solving to resolve any discrepancies in your head.

Everyone uses the same method of trial and error to refine their processes further. The feedback you receive helps you develop a strong basis for your theory. And, despite living a life very similar to someone else's, you'll come to conclusions differently.

But, no matter what anyone else tells you, it doesn't matter what structure this coherent system takes. For some people, their minds look like perfectly lit hallways filled with cabinets and binders that perfectly lead to their conclusions. For others, their minds are filled with rows and rows of filing cabinets, some turned over on their sides with paper spilling out, some standing upright with their drawers flung wide open, and some upside down and on the ceiling. The messiness of the system doesn't really matter to those people. As long as their inner clerk can rifle through all of the papers and know exactly where they are, the system makes sense.

The Perspective Shift

It's surprising just how many theories you develop in life. What's perhaps even more astonishing is how many of those perspectives actually change.

When you're young, you assume you know so much. And, between the ages of two and twelve, because you have experienced so much cognitive growth, it's hard to believe that you're wrong about much. But the limited knowledge you have in your early years, tested though it may be in your short-term experience, doesn't always pan out the way you think it will.

Let's take another deep dive into your six-year-old mind.

Remember how at that age you believed that because sugar tasted good, it must be good for you? At your house, your mother significantly limited the amount of sugar you consumed. Every time you reached for that sugary cereal at the grocery store, you received a stern warning to put it back. After a while, your only recourse was to stare at it longingly as you passed through the aisles.

But your best friend down the street was allowed far more sugar than you. When you stayed for a sleepover at his house, you were greeted by marshmallows, Rice Krispies treats, and chocolate milk in the mornings. To your young mind, it was truly a dream.

When you went back home, everything was different. The unfrosted Wheaties your mother gave you seemed to taste even more like cardboard than ever. You knew there was no point in begging your mother for chocolate milk, but it didn't stop you from glaring around angrily before you left the house. To you, your mother was truly a dictator. Your friend's mother loved him far more than your own mother loved you. After all, if she truly loved you, wouldn't she give you all those sweet treats as well?

Now let's skip forward several years. When you leave college, you are in great condition. You've engaged in sports throughout your academic career and maintained the same diet that your mother did when you lived under her roof. By happenstance, you run into your old friend from down the street. He's now at an unhealthy weight and carries insulin around with him to combat his type 2 diabetes. It's startling to see how much has changed. Maybe your mother was a good parent after all.

So, what happened?

The circumstances didn't change, but your perspective did. At a young age, your limited experience couldn't account for the possible damage that might come from eating too much sugar. While you once believed that your mother was a bad parent, looking back now, you realize just how good a parent she really was. As your scope grew, your axiom changed.

This type of perspective change happens throughout life. You might realize at age forty just how stupid your perspectives were at age twenty. You might see the flaws in your parenting style at age eighty when your own child is realizing those same problems as a grandparent at age sixty. It's a never-ending cycle, and it's supposed to be. You're just not living life if your perspectives never change.

Circumstances and experiences always change. The only way humanity can make sense of life is by testing limits to create a self-consistent system. Life is a never-ending cycle of repeated tests that only ends with the end of life itself.

Chapter 2 Questions

- *What examples do you have of developing some of your self-consistent systems? What methods do you use to develop them?*

- *What are some of the methods you use now to develop a self-consistent system that you didn't use a few years ago?*

- *What missteps have you seen in your or someone else's life that have led to a limitation of scope or a flaw in judgment?*

CHAPTER 3
Scientific Methodology to Make Sense of Theories

As we've already established, understanding life comes down to testing our theories. It's one of the best processes to test data and expand knowledge. And although you may not have labeled it as such, you've been using the scientific method to make sense of your theories for your entire life.

Think about the last time you made a major decision. Maybe you were looking for the right spouse, buying a house, or raising a child. How did you settle on a conclusion? Did it come easily to you? Or, like most decisions, did it take a lot of trial and error? Well, if you are one of the people who got it right the first time, you can count yourself among the lucky few. The rest of us had to go through a series of failures and successes to find the best solution.

Most of the time, it's the failures that give us the most knowledge. How would we know how things could go so right if we're not aware of what could go so wrong?

This series of trial and error is what has propelled humanity from caves to artificial intelligence. In fact, it's so important to the way that humanity has developed—and it's been practiced across cultures— that it's been named and taught for centuries. You might know it as the scientific method.

So, how can we actually make sense of our surroundings? Well, it takes one step at a time and one theory resolution at a time.

The Will to Fit In

Let's think back to when you probably first heard about the scientific method. You might have been sitting in an uncomfortable seat at a cheap fake desk, staring up at your teacher as he drones on about how humans come up with scientific explanations. Of course, all you've been thinking about all day is that red zit at the tip of your nose.

As you glance to the side, you notice a classmate trying to pick breakfast out of her braces. You try not to look behind you at the cool kids who are sitting in the back row. But, when you do finally pluck up the courage to give a sidelong glance behind you, you notice a cheerleader's smirk. It's all you can do to throw a hand over your nose.

When it's finally time for lunch, you head out to the lawn to sit down with some of your friends. They're new. It seems like, ever since middle school, you haven't really known where you fit in. You're sticking with this group of friends because you just kind of fell into their circle. As you start opening your homemade sandwich, you hear someone comment, "That Miss McCaffrey is such an idiot." Everyone in the group starts laughing and nodding their heads. You go along with it, slightly smiling as you think, *I actually really like her.*

Just as the bell is about to ring, sending you back to class, you're pulled aside by Tim, one of the more popular boys in the group. "I just wanted to let you know," he says, "some of us were noticing that crochet scarf you have and …" He grimaces. You look down at the scarf that your mother gave you, and color immediately rises in your cheeks. When you look up at him again, you give a nonchalant nod before taking it off of your shoulders and throwing it in the nearest garbage can.

The next day, you're dressed in more muted colors. When you show up to Miss McCaffrey's class, you pull an oversized jacket closer around you. You take a seat in the back near some of the other cool kids. And, instead of paying attention, you start doodling in your notebook. *Maybe I was completely off base,* you think. It makes more sense to blend in with the cool kids than to wear that scarf that hurts your heart a little every time you think about it in the garbage can.

Alright, so not everyone goes through the same annoying, frustrating, infuriating, or embarrassing experiences in school. And yet, everyone has experienced the desire to fit in somewhere. As a teenager, it was one of the first times that you truly felt like you could express yourself outside of your family. It's a time of self-realization, and you start to develop a better knowledge of yourself through observation. There were people you looked up to and people you despised. Through that observation, you developed a theory about yourself.

Throughout your teenage years, you developed a theory about yourself that constantly went through alterations. You were plagued by that dreaded question: Who am I?

Your teenage years were a time of trial and error. Some weeks, you didn't know who you would be from one day to the next. Your beliefs, thoughts, and actions were dictated by the perceptions of others as you struggled to find yourself. Many followed along with people whom they believed had found themselves already. Theories were tested and conclusions drawn.

By the time you reached the end of your teenage years, you were not so much a product of what you learned in books but an amalgamation of what you'd seen from other people and what you believed yourself to be.

The Leap Forward

The will to move forward, the intense desire, comes from the desperation to learn more. Humanity has crawled through various ups and downs to find solutions to questions, from the simplest to the most complex.

And yet, all these solutions came by the same means.

Perhaps one of the most well-known examples lies in Einstein. Today, he is known as the visionary responsible for relativity and dipping into the quantum realms, a position that seems almost untouchable. But it all came from a simple thought experiment.

Decades earlier, in 1887, the scientist Heinrich Hertz experimented using a light source, a metal plate, a vacuum tube, a positively charged plate inside the vacuum tube, and an instrument to measure the voltage and current created by electrons. By shining different colors of light onto a metal plate inside a vacuum tube, electrons are ejected from the metal surface. Low-frequency colors (red) result in no emitted electrons, regardless of how bright the light is. When you switch to a higher frequency light (violet), electrons are ejected from the metal surface. Hertz measured the electrons ejected when they were attracted to a positively charged anode, which created a flow of electric current.[9]

---

9   Michael Fowler, "The Photoelectric Effect," *University of Virginia Physics Department,* *https://galileo.phys.virginia.edu/classes/252/photoelectric_effect.html.*

The experiment ultimately showed that electrons moved with greater motion when there was a higher frequency of light. Hertz was at a loss to explain the phenomenon based on classical mechanics. It simply didn't make sense.

Looking back at theories established in the past, Einstein started asking questions about this light phenomenon.[10] By classical standards, if you make the light brighter, you give more energy to electrons. Also using these theories, you should be able to predict that electrons will eject from the metal if any color of light or any length of frequency shines on it.

But, if classical theories couldn't come up with a realistic explanation, it was only logical that something outside of those theories would solve the problem.

The revelation came as a new way of thinking about light. Instead of light moving in waves, perhaps light was also tiny packets of energy. Each photon would carry energy depending on the light's frequency or color. Higher frequency meant more energy, and lower frequency meant less energy per photon.

This new concept of light would affect electrons differently from the classical model. When a photon hit the metal, instead of emitting the same amount of energy, regardless of frequency, the photon would give its energy to an electron. The more energy that the photon had, the more power it had to overcome the force holding the electron in the metal.

Einstein's theories were purely theoretical: He didn't conduct the physical experiments himself. He was far more of a theorist, really. Yet,

---

10   Albert Einstein, "On a Heuristic Point of View Concerning the Production and Transformation of Light," in *The Collected Papers of Albert Einstein—The Swiss Years: Writings, 1900–1909*, vol. 2, English translation supplement (Princeton University Press, 1989), 86–103. Einstein's questions about light and photon theory and his work on electrons completely altered classical theories of physics.

it prompted other physicists to test his theories. A contemporary of Einstein's, Robert Millikan, experimented with the theory for over ten years before eventually validating Einstein's photon theory.[11] The foundational ideas laid the groundwork for future experimentation and theories regarding light and the subatomic.

Einstein's work in the quantum realm was the pure application of the scientific method. Careful observation and hypothesis formation, followed by rigorous analysis, revolutionized the way people thought about the smallest particles. He was only one of many who dove into particle analysis, but his contributions represent some of the foundational elements of quantum theory today. His work on the photoelectric effect alone earned him the Nobel Prize in Physics in 1921, but it's far from the only thing that promoted a deeper look into particles.

Where We Go from Here

This phenomenal take on science developed because many people had the persistence to look beyond what was already known. And that's all science is: the development of theories built on each other, each involving extensive experimentation and subsequent development of new theories. Isaac Newton's laws, Albert Einstein's theories of relativity, and Stephen Hawking's explanations of black holes—they're all explanations built from the series of experiments and observations of past thinkers.

Looking back, it's sometimes hard to reconcile that five thousand years of written human history have only now led to the conclusions that we currently see. Shouldn't there have been more scientific explanations by now? Logically speaking, the past hundred years have

---

11 Gerald Holton, "Quantum Milestones, 1916: Millikan's Measurement of Planck's Constant," *Physics*, April 22, 1999, *https://physics.aps.org/story/v3/st23*. Though he didn't accept the results for many years, Millikan effectively proved the reality of photons.

yielded an explosive amount of new data in technology, so it only makes sense that, if such brilliant scientists had always existed, we would be far more technologically advanced than we are now, right?

Well, it's tricky.

Some tools and measurements available to those in the past were enough to come up with some revolutionary conclusions. For example, Eratosthenes of Cyrene calculated the approximate circumference of the earth in 240 BC. He observed the sun's angles at two different locations in Egypt, one in Syene and one in Alexandria. The sun at noon on the summer solstice was directly overhead at Syene, evidenced by the fact that it shone straight down a well. On the same day and time, the angle of the shadow of an obelisk in Alexandria was slightly different. Calculating the distance between the two locations and the angle difference, Eratosthenes estimated the world's circumference to be about 24,662 miles.[12] Remarkably, he was only short by 239 miles.

On the other hand, though the term *atom* was first coined in ancient Greece around 400 BC, actual evidence of the atom's existence and a direct understanding of what it was didn't come about until the twentieth century. There were significant limitations that prevented scientists from seeing them up close. It wasn't until the electron microscope was developed that people could actually see the subatomic.

In short, scope limitations existed far before many theories had experimental evidence.

Skirting the Boundary

It's possible to have foresight into the limitations of our theories by understanding the scope. First, like Einstein, we need to identify our theories' boundary conditions and initial conditions.

---

12   *Britannica*, "Eratosthenes," last updated December 15, 2020, *https://www.britannica. com/biography/Eratosthenes.*

Boundary conditions describe the limitations within which a theory can thrive while reflecting the scope of interest to be known or learned. A good way to think about this is illustrated with a vibrating string. Let's imagine that both ends of the string are tied off and don't move. When a vibration is sent through the string, only the length of the string vibrates. We can measure the amplitude and frequency of the vibrations on the length of the string while our boundary conditions keep the string fixed. In the same vein, the initial conditions break down what we already know about the string. First off, we know that the movement of the string at the boundaries is always zero, since the ends are locked in place, but we can set initial conditions where the vibration starts by picking a length relative to the fixed point. We can also describe how fast the string will move at the point of the initial vibration. In essence, by understanding the boundary conditions and initial conditions of a theory, we can establish what we already know.

The next step in establishing a theory is the process of making assumptions. In the case of mathematics as a tool for science, these assumptions are usually called axioms, which are considered to be true without proof. In social science, we usually make assumptions to start with, as we do in real life while we are trying to make sense of something. The truth is, we can't find conclusions to our theories if we aren't willing to guess a little. Luckily, we have thousands of years of written knowledge to base our ideas on. We can usually create educated guesses based on our own prior experience and the experience of others.

When establishing a theory, we need to understand where the boundaries lie. Something as simple as an apple dropping from a tree to the ground has its own boundaries. The tree is the apple's origination point, and the ground is its final destination. And they are both on the earth. As for the initial conditions, we know that the apple started at rest, and it ended at rest. We also know that the apple gained speed as it tumbled toward the earth. It's that guess of what happens in between that's the foundation for a scientific theory.

So, what exactly happens between the time of the apple's fall and its connection with the ground?

We can establish that any apple, regardless of its location on the tree, will fall to the ground at the same rate of acceleration. We can measure the acceleration using an accelerometer to find that each apple falls at roughly the same rate of 9.8 meters per second squared (m/s^2).

We can easily test the boundaries for this phenomenon. What happens, for example, if we drop that same apple from the top of the Eiffel Tower? Will it have the same rate of acceleration? Or, what if we change the type of object falling from a tree, using a bowling ball instead of an apple? Will it yield the same results? We can determine from the results of all these experiments that we will achieve the same acceleration every time. Using language to interpret the data, we can assign this phenomenon the name *gravity*. And now we also know that gravity may happen on any planet besides the earth, but the rate of gravity will no longer necessarily be 9.8 m/s^2.

Of course, if you've been paying attention, this process looks familiar. It's the same old story: Develop a theory, perform an experiment, come to a conclusion, and repeat the process. But what if you're reaching much further into the unknown?

Challenging Limitations

Scientists such as Einstein developed their theories by pushing the boundaries of current knowledge. Einstein had studied physics and mathematics in depth and was heavily influenced by previous scientists during his tenure at universities. In particular, the Michelson–Morley experiment of 1887, designed to detect the presence of an ether (the substance then believed to be the one that permeated space and carried light waves), suggested that there was no such medium in space. In other words, there was nothing in space that slowed down the speed of light.

The idea was revolutionary, and it sparked Einstein's interest.

He developed two axioms in 1905 to explain this phenomenon: (1) the laws of physics are the same for all observers who move at the same velocity, and (2) the speed of light is constant for all observers in a vacuum.

As Einstein pondered Newton's theory of gravity—namely, that gravity is a force that pulls objects toward each other—he started to question how gravity would act across space. He developed the equivalence principle to describe gravity slightly differently.

Think about it like this: Imagine you're in a closed capsule in space. You're far away from any planet or star, really anything that could be pulling you gravitationally. If you were to accelerate at a rapid pace (let's say 9.8 m/s^2), you'd feel yourself being pulled to the floor. It's the same sensation you'd feel sitting at the surface of the earth. You would not feel the same gravitational force in both instances, but you'd feel the same effects.

So, Einstein reasoned, what if gravity could shape space and time? He developed a new theory that suggested space-time moved like fabric. Massive objects with intense gravitational fields could warp the fabric, actually bending space-time around heavy objects. Any objects, such as planets, that moved around these intensely heavy elements would move along the curves at the boundaries of the space-time distortion.

The basis for these new theories on how gravity interacts with space-time laid the foundation for some of the most impactful techno-logical and scientific advancements in the twentieth century. Because people started to think of gravity as not just an invisible force but a geometric property of space-time, many began to ask questions about the formation of the universe. New areas for research opened to account for the types of natural forces. This began with thinking of unique astrophysical concepts, influencing the study of black holes and the expansion of the universe. Technologies, such as GPS, were designed to account for the relativistic effects of gravity to provide precise data.

But perhaps the most interesting part of this whole story is Einstein's reaction to his newly created framework of general relativity. Yes, his theories would certainly answer questions that couldn't be solved with classical views on gravity, but were they complete?

This rather broad overview of general relativity that we've discussed may make it appear that relativity is an easy concept to understand, but it's something that Einstein struggled with for the rest of his life. General relativity is mathematically complex. It requires a deep understanding of a variety of mathematics, and it's something that Einstein was concerned the scientific community wouldn't understand. To combat the inconsistencies, he engrossed himself in long hours poring over mathematical concepts. He also involved fellow mathematicians, such as Marcel Grossmann, to handle the geometric aspects of general relativity.

In addition, he wondered how it would fit in with quantum mechanics. General relativity was good at explaining large-scale gravitational phenomena, but he wasn't sure it would mix well with the microscopic world. His famous quote, "God does not play dice with the universe," came in response to his fear of the probability of quantum mechanics. It was something he struggled with for the rest of his life.

Combating Inconsistencies

If you're not a scientific person, a lot of what Einstein did might seem like it goes right over your head. However, there are some things that we can learn from this rather extreme example, such as how to deal with inconsistencies.

As mentioned earlier, trial and error is the only way to develop a sound theory. Most of the attempts will fail. But to find the best possible conclusion to your theories, you need to identify their inconsistencies. Those inconsistencies are flaws, and they invalidate your theory.

It's hard to see it from an outside perspective, but those flaws and inconsistencies are among your most valuable assets when developing new theories. You know where you went wrong. And, if you know where you went wrong, you can make changes. It's possible that you simply need to expand your scope. Just like with Einstein, when you expand or limit your scope to a degree that makes sense, you might open brand-new avenues.

We base these new realities and understandings of limitations on our criteria. The first question we asked was, "Is this consistent with reality?" The older we grow, the more in tune with those criteria we become. You'll become more aware of yourself and what is or isn't impossible. Once again, it comes down to pushing boundaries, but within the scope of reality.

You'll start to develop criteria based on a value system: what you believe to be right or wrong. Your personal evolution is based solely on what you value. And, correspondingly, your final judgment is based on that key fact. As you grow older, your criteria and value systems change as your scope changes. You discover new facts, include more people in your life, and ultimately change your perception of reality.

The Refining Process

Let's take a second to step back and absorb everything involving the scientific method in our personal lives. It's a lot like that bit of clay you played with as a child. You might have created a mug for your mother, but it was lumpy and didn't hold its shape. Maybe you crafted a dog that, years later, you can't quite make out what it is. With the benefit of hindsight, you can definitely see some flaws.

As it turns out, life is a lot like that. What we heralded as something untouchably spectacular when we were children, we now look back on with a slight grimace. Most of the time, it's not something that we consciously think about. I mean, you thought you were presenting a sculpture akin to one by Michelangelo when you were a child. Now, however, your perceptions have definitely changed.

Oftentimes, it's difficult to keep track of these ever-evolving criteria. It makes sense in theory to look back and notice significant changes, but how do you really know when you're evolving?

Luckily, there are ways to know for sure.

Establishing Metrics

For the entirety of the existence of human written language, we know that people have established metrics to better understand evolution. Some of that was as simple as establishing the first numbers, written in clay and passed from person to person. As civilizations progressed, they were able to look back and see where they had been.

When establishing metrics for your own evaluation, the first important step is to set up indicators that will assess your performance in relation to those criteria. Asking yourself questions such as, "Did I reach the outcome I thought I would?" helps you become more self-aware of where you started and where you ended.

Doing this for every decision—every internal thought—can seem excessive. It can start feeling like your life is just a series of theories and conclusions, which can take the joy out of living. So, instead, include the measurements for your objectives and comparisons. On a broad scale, are you reaching your goals? What significant roadblocks have prevented you from establishing the right criteria? Are you establishing the theories that will help you to get closer to your goals?

Gathering and Analyzing Data

After you've gathered the metrics that assessed your performance or the outcomes of your theories, take a moment to think about them. Does the data support your theory?

This is where a lot of people get tripped up. There's a sense of pride in establishing theories about life. If you've gone your entire life believing one thing—confined to a limited scope—only to realize that you were completely wrong, it'll sting a little. That's completely natural. It's never pleasant to find out that you've been wrong, but it can be pleasant for you to know that you see the changes you can make for the better. And it allows you to engage in some of life's most important processes.

Refining Theories

After you have established theories, experimented, come to conclusions, and critiqued your criteria for establishing those theories, you can finally refine them. You might find yourself adjusting your theories based on scope: Maybe you thought a little too narrowly as a child and now have outgrown some of those initial thoughts. On the other hand, maybe what you value has changed: Maybe you realized you needed to change the way you thought about the world after you had a child.

Whatever the reason, you're constantly adjusting and refining your theories. Using the scientific method in life allows for constant change. You're constantly revising your position based on new information, new scopes, or new changes in values. You won't always stay the same, because your perception constantly changes. This actually means that you are growing. And you continue to grow as long as you keep refining your theories.

Along with your growing process, making improvements to your theories and criteria becomes a natural ritual. You'll have learned over the years that the scientific process is fluid, and you've incorporated that fluidity into your own life.

The Teenage Evolution

But let's take a step back one more time, back to the time when you constantly critiqued yourself and changed your perceptions of life seemingly every minute.

Once again, you're trying to fit in with the group. You might have started to take a long, hard look at what nutrition is doing to your figure. As a child, you believed that anything that tasted good was good for you, but as a teenager, you're starting to realize it has a detrimental effect on your health. Maybe you're trying to slim down or even bulk up, if you haven't cared about your health that much yet. Either way, that sleeve of OREOs isn't helping.

As you start to see more significant changes, you take it to another level. The endless barrage of sugar stops at the door. Instead, you're all about protein shakes and kale smoothies. Sometimes, it's hard to stomach, but it's worth it to see the changes.

But after a while, it goes a little too far. You start to become ill and faint throughout the day. While you used to be able to run five miles a day, that is becoming more difficult all the time. Maybe you finally show up at the doctor's office to find out what's really going wrong, only to discover you've taken it a step too far. All that time trying to fit in has made you reduce your calorie intake to around a thousand per day, far below what you need.

Maybe it's not the "tale as old as time" that everyone experiences, but it's a common thread seen among many in their adolescent years. Peer pressure and internal perspectives change significantly in those few years. Your perception of self modifies nearly daily.

As you get older, you may start to care less about what people think, but your self-perception continues to evolve. You start on a path of self-discovery. The more you experiment with the theories you've developed over time at any age, the more logical conclusions you'll create. It's the scientific method played out in real time, and it's always changing.

Chapter 3 Questions

- *How have your perceptions changed since adolescence?*

- *What series of problem-solving methods have you developed from years of trial and error?*

- *What metrics have you used to evaluate your performance or outcomes in your life, and what value system have you developed that forms the basis of your criteria for improvement?*

PART II:
Making Sense of Life as an Individual

CHAPTER 4

My Life, My Experience

Making sense of nature is a human-wide endeavor. It's taken the theories of countless brilliant people to produce the basis of what we know about nature today—every theory, every experiment, and every conclusion stacked up over time like building blocks, growing with every unique theory and perspective. New rationales for old queries have yielded answers to some of life's most difficult questions.

In a way, those answers were easy. Sure, they were hard-won, but with the perspectives of so many people striving to answer the same questions, there was a lot of help along the way. There were rights and wrongs, established truths and uncovered falsehoods. Though there are still some disputes about conclusions drawn, a lot of things are basically black and white, especially when the scope is well understood or well defined.

It gets a lot more complicated on the individual level.

Philosophers have attempted to answer some of life's most complex questions since the beginning of time. Each region, country, or society has its own suggested road map. Nearly everyone has had theories on how to live the best life. But, since everyone is different, how can you create a one-size-fits-all solution?

Understanding who you are as an individual requires a lot of complex thought. We're not animals, which gives us the benefit of self-actualization, but understanding our true natures comes with a heavy burden. Blessing or curse, we bear the responsibility of living morally.

It's fundamentally important to establish our values and hold them up to the standards of society, but the analysis process is difficult. How do you actually know how to make sense of your life? Is it through the lens of other people? Is it through solitary introspection? Is it a mix of the two?

Luckily, each of us was born with the desire to find out more about ourselves. That innate craving to learn more about nature helps us better understand who we are. The scientific method, commonly used for objective viewing of reality, holds the key to understanding ourselves.

So, the question becomes, How can we make a difference in our lives when we've just gotten ahold of nature?

The Decade of Experience

Imagine yourself in the front seat of your car, driving away from the only place you've ever known. You've started a new job, and it's taking you far away from your family. You take a deep breath, throw your shoulders back, and tell yourself you're doing the right thing. You've been dreaming of this moment for a long time. Being on your own always seemed so magical, especially if you didn't have to deal with your younger brother. You have your own car, you're on your way to your own apartment, and you're only taking a few things with you.

Your mom kissed you on the head as you headed out. You knew it would be the last time for a while.

Let's skip forward a bit. A few years after departing from your home, much wiser from years of work experience, you're starting to notice a shift. Your girlfriend of the past year is getting a lot more serious, and you want to pop the question. It seems like a big step, but you're ready for it.

You try not to hold your breath as you walk down the steps of your apartment and open the door to the city outside. You've had everything set up for weeks. You've reserved the park and have already laid down a blanket and a bottle of champagne. Your best friend is hiding in the bushes with a camera. You've truly thought of everything.

But as you guide your girlfriend toward your reserved place at the park, you notice something's a bit off. She's not smiling as much as she usually does. She seems distracted, looking around her, at her shoes, at the birds flying overhead, really anywhere but at you. *She's probably been wise to me this whole time*, you think to yourself.

As you sit down, you notice her fidgeting. But you're fidgeting, too. You've got the ring, and you're trying as hard as you can to keep her from looking at that square box in your pocket. Finally, as you hoist yourself up onto your knees, you look into her eyes and say, "Will you—"

"I think we should break up."

Ah, your twenties. It's that decade in life when you really start to experiment. You're finally out on your own and doing things you've never done before. It's the point in life that you dreamed about since you were a child. Independence, success, and personal growth. It all happens in your twenties.

But, behind the glitz and glam that comes with your first taste of independence, there's fear. The relationships you develop, whether good or bad, profoundly influence your life. While you'll always have your family, your twenties also bring about the reality of doing things on your own. Instead of a walk down the hall, a phone call might have to suffice. Your best friend in high school may turn out not to be the person you thought they were. Life is just much rougher than it had been a mere two years earlier.

Early adulthood marks the end of rapid cognitive growth. Your brain finally stops developing around the age of twenty-five. Up until that point, you were constantly learning, constantly making cognitive connections as you grew and learned. After your mid-twenties, life is much more about the development of personhood. It's the point at which individuality develops. The resilience that comes through your response to challenges and setbacks is completely unique. You are finally becoming your own person.

Human Potential

The sudden shift in experience you feel when thrust into a new situation prepares you for analyzing situations and environments outside of your control. It's the point at which you start progressing into the abstract. It's the starting place for personal development and understanding who you truly are.

It's funny how many people really see a significant shift in their twenties. Among the most famous is Friedrich Nietzsche.

Nietzsche truly began exploring answers to some of life's most mysterious questions at the age of twenty-four, when he became a professor of classical philology. It was quite an astounding accomplishment, as he still holds the record as the youngest person ever to hold that position.[13]

---

13   R. Lanier Anderson, "Friedrich Nietzsche," in *The Stanford Encyclopedia of Philosophy*, ed. Edward N. Zalta and Uri Nodelman, March 17, 2017, *https://plato.stanford. edu/entries/nietzsche/.* This provides more information regarding Nietzsche's past.

Nietzsche's skyrocketing rise to the highest intellectual milestones was a blessing and a curse. Just a few years later, he began experiencing severe health issues, including migraines, eyesight problems, and a variety of other physical ailments. He went into hiding, preferring prolonged periods of solitude and introspection to company.

Eventually, it forced him to seek nonconventional paths. His internal struggles made him intimately aware of his own limitations. It became an intrinsic need to find a solution, to ultimately master his problems. His philosophy became much more focused on individual strength and resilience. He coined "the will to power" as one of his philosophy's core concepts.[14] It emphasized each individual's desire to assert and enhance their own influences or themselves and their circumstances.

As a result, he began to question the current norms and truths. Suddenly, traditional values and moralities didn't seem to satisfy the necessities of mortality. He found that these limitations often suppressed individual potential and creativity. Instead, he posited, challenging prevailing philosophical doctrines became more important than ever. His view of reality became much more fluid, dynamic. What was right to some wasn't right to others.

For the rest of his life, he felt the need to focus on the potential for human greatness. He emphasized the world of power, rejecting laziness and encouraging everyone to constantly challenge themselves. With enough willpower, anyone could achieve extraordinary things.[15] Personal development was a continuous process that

---

14   Bernard Reginster, *The Affirmation of Life: Nietzsche on Overcoming Nihilism* (Harvard University Press, 2006). Nietzsche was notorious for critiquing traditional moral values and exploring concepts beyond truth, morality, or the nature of humanity. He challenged the idea of good versus evil in favor of seeking underlying human behavior.

15   Jonathan Mitchell, "Nietzschean Self-Overcoming," *Journal of Nietzsche Studies* 47, no. 3 (2016): 323–50, *https://www.jstor.org/stable/10.5325/jnietstud.47.3.0323*. Self-overcoming is the way to evaluate your ethics.

required continual self-overcoming and transformation unburdened by external limitations or traditional morals.

The Theory of Life

We understand the world through what we experience. We can predict the way the seasons will behave based on what we've seen. We can find solutions to problems based on how we've solved them in the past. In essence, our experiences dictate how we interact with the world and help us understand it.

But it gets complicated when life becomes unpredictable.

That car that swerved into your lane and almost crashed into you is something you didn't expect because, in your experience, drivers mostly stay in their lanes on the road. That bird that flew into your office window scared you half to death because, in your experience, most birds don't fly into buildings.

Though you base your understanding of life on the predictability of outcomes, there will always be things that you just can't anticipate. There are an infinite number of possibilities that come from every decision made by you or anyone else. And because you exist for decades, you'll experience countless variations of your experiences.

As you try to make sense of these variations, you'll keep asking *why*.

For example, if you missed your bus going to work even though you arrived at the bus stop on time, your mind will search for rationalizations. Maybe you check to see if your clock was slow and you didn't realize it in time. Maybe the bus left early from its station and arrived at the stop much earlier than it normally does. Maybe you were on time but the passengers that morning boarded the bus faster than they normally did. Maybe, maybe, maybe.

The infinite number of explanations for the variety of experiences you have in life can drive you insane if you let it. Getting stuck in a never-ending loop of asking yourself *why* is infuriating.

But human nature has a solution for that confusing and frustrating cycle: Just stop asking. Eventually, there's no point in continuing to rationalize your experience, sometimes simply because you cannot afford more time to think about it. The possibilities you come up with will only become less and less likely, and you'll ultimately be unsatisfied with whatever is left, because it doesn't help you serve your purpose anymore. At the outer reaches of the explanations for your experiences, there is only so much value you can get from continuing to press to serve your purpose.

Even though there is an infinite number of possible outcomes, there is a point at which we naturally must stop. There is only so much that we understand through our experiences because we have only experienced a limited number of possible solutions.

Expanding the Scope

It's at this acceptance of our limitations that we can truly start to understand the scope outside of ourselves. Though *you* can't experience everything in the world, you can learn about others' experiences. Even if you never saw Marie Curie's discovery of radon, you can read about it. Even if you never saw Mozart creating *Don Giovanni*, you can still access the manuscripts, letters, and personal accounts of the people around him. Though your personal scope is limited, it doesn't mean that you can't begin to understand others' perspectives through their own eyes.

Your experiences, whether passive or interactive, and the generated results form the basis for your understanding of your surroundings. The solutions to problems in your life, solved by the beliefs of your own axioms, may seem to you like the perfect explanation for someone else as well. After all, if you've established that something is universally correct for you, it only makes sense to you that it should be universally correct for someone else. At least, that's what you think.

But something altogether unique happens when you start adding other people to your process of making sense of life. The universal beliefs that just make sense for you are only universal to you, as your axioms may only be assumptions when applied to someone else. For you, it might appear that everyone's parents are pushing them to go to Harvard. But, across town, there could be a child whose parents expect him to start working right out of high school.

Though there are some axioms used by science in a scope universal to all of us, unique perspectives make finding the beliefs much more complicated in our lives. When you don't know if something is true or not, you make assumptions so that you can develop theories. The only way to find out if these assumptions are true under the given scope is to test them across multiple perspectives within the given scope, but your own limitations make it impossible to test every possible outcome as scientists try to do. It's at this point that you accept your limitations and create a basis of understanding of society based on the assumptions you've created.

Though it may not seem like it, you're using math and logic together with your language(s) to make sense of life. In your head, you've created a theory, experimented, gathered results, and come to a conclusion based on your findings. Your entire theory on life is based on information gathering like an experimental physicist—basically, your experience. The process of developing a theory, testing it through new experiences, and developing a prediction for its outcome is the basis for human existence.

But creating theories and testing them through experience on a wide scale is much more difficult. As your scope changes, everything else changes, too. Axioms change at different scope sizes.

When you're ready to start expanding your scope to understand more about the world outside of what only you experience, you need to make theories about things you haven't yet experienced. It's at this point that you start making widespread assumptions to fill the gaps that you can't experience or don't understand. You may learn that, while you feel at home in a small apartment in New York City, your cousin, who comes from the country, feels cramped and uncomfortable. You might assume that everyone feels the same way you do, but when you meet your cousin, you find your assumption is wrong.

The problem isn't necessarily your axiom but your scope. The axiom you created based on your understanding of your environment isn't necessarily wrong when it's confined to a small area. Other people who live in New York City, and specifically people in your apartment building, may also feel that it's cozy to live in a small apartment. Even though your experiences are different from theirs, you all have developed a similar perspective or the same axiom.

But inviting your cousin from rural Texas has significantly expanded your scope. You are no longer talking to someone with a similar background and understanding of the world. The experiences that your cousin has gone through are entirely different from your own. By expanding your scope, you've found that your assumption that everyone thinks a small apartment is cozy was incorrect. After you conclude your axiom was incorrect under the expanded scope, even though it might be correct under the original smaller scope, you can adjust it to fit a new understanding.

Humankind is programmed to experience new things. But the process of developing axioms and testing assumptions through experience can be painful work. It requires openness, which isn't comfortable. And yet, it's the only way for our understanding to continue to grow.

To enjoy happiness, we can try to expand our understanding daily. We crave progression, either personally or as a species. It's in our nature to try new things and continually expand our knowledge. You'll see that when someone is unhappy, they may sit around and do nothing all day, living in a confined scope. We crave a challenge, and when we are stuck in a small, progression-less section of our lives, we become much unhappier. The more uncomfortable we make ourselves, the happier we may be. After all, discomfort doesn't necessarily equal unhappiness. Just imagine your first date!

We're naturally designed to develop a theory of our own life. We develop a desire to learn more, yearn for future experiences we might never have considered before we gained more experiences. It's a never-ending cycle that we all share, but what we desire is unique to each individual.

Developing an Individualistic Experience

"But," you might ask, "are there truly unique perspectives to be had for myself as an individual? Aren't all philosophies on life derived from someone else's?"

Well, there seems to be some truth to that. The foundations of thought are often developed by the perspectives of those who have gone before us and experienced things similar to what we have. But no one has your unique experience in life and, therefore, your unique perspective on life. No one's experiences are exactly the same as yours. Some people have experienced the same things you have, but the results were completely different. Some people, worlds away, have experienced things you could never imagine. Even twins experience life somewhat differently from each other.

Let's return to our Nietzsche example. His unique perspectives on life and the development of his own theories came about because of the trigger of poor health. Who knows if he would have continued with the same philosophical approach to life as all other scholars of philosophy at the time if he hadn't had physical ailments? His philosophical ideations at the time were quite unique, which inspired the philosophies of others.

Unfortunately, many people don't believe that their philosophies are worth mentioning. Most people who feel they don't have a unique philosophy on life get stuck on the written aspect. Some of the greatest philosophers of all time wrote down what they thought of their experiences cohesively. They say to write clearly is to think clearly, and if you struggle with writing things down, you might start to question your theory's validity.

While there is truth in developing set criteria to define your theories on life, just because they're not written in a book doesn't mean your theories don't exist. They're merely evolving.

The creation of personal meaning is a dynamic process. It involves the development of interpretation of thoughts, experiences, and emotions into meaningful insights. Your development of identity—who you believe you are based on your ancestors, your childhood, your positions in life, etc.—guides your personal philosophies. To some, it comes down to genetic makeup: Some blame their hot-tempered Italian heritage for their bold perspectives on life, while others claim their long line of Buddhist history to explain their inner peace and tranquility. Others believe their philosophical development comes down to experiences in their own lives.

Whatever the case, you'll often find that these identities change over time. At different points in life, we all develop our own sense of our personal identities. And they do evolve along the paths of our lives while we're experiencing more.

The Autonomy of Learning

Often, these chances to learn more about yourself, to identify who you really are, go unnoticed. I mean, when was the last time you told yourself, "That summer just flew by!" It's easy to be unintentional about life. It's easy to say, "I'll be happy when …" I know it's cliché, but understanding the journey through life and stopping to smell the roses is one of the best ways to make your life more intentional.

Learning who we are is a dynamic process that forces us to engage with our environments. If we're not engaging with those environments—in short, if we're not actively analyzing our perspectives and experiences—we're not growing. Every day, we're actively creating a system of self-growth and self-discovery. Unfortunately, if we're not actively analyzing our perspectives on what's happening to us, our growth slows significantly.

It's a common belief that we change solely based on other people we encounter. There is something to that. Our identities often change based on important people in our lives. Our relationships twist and turn with the tide as we sail to a better understanding of who we are. It's an important part of our personal development to interact with a variety of people who shape who we are.

And yet, the only way that we essentially grow is through our own autonomous learning. The way we react and change based on what we've experienced is completely unique. We form information and question beliefs based on our desires and perceptions.

The acceptance of our autonomous learning is key to developing a sustainable path forward.

Defining Real-World Progression

But, just like when you find a limit to your questions, when you decide an infinite number of possibilities are impossible for you to figure out, you need to find where the limitations are to your scope. Your scope is limited to what you can physically see, literally and figuratively.

So, how do you consciously create axioms to reliably guide your life? As we've already established, it's by adapting your axioms over time and considering the scope and the limitations of your experiences, which depends on a combination of math and logic, trial and error, theory, and application. But conscious axiom integration requires the use of language. Your definitions of your axioms bring them into focus.

The language you use to describe what you see in the world is again limited to what you know. Language is a tool, but it's limited by your scope. If you don't see new things or hear new words, your axioms can only exist in a limited scope without expansion.

Once you develop those axioms in words, you can compare what you've experienced.

People like to think that they've figured out the nuances of the world, but those axioms they've developed don't necessarily align with what can be. It may be hard to accept that a loved one has cancer, but it doesn't change the reality. As a child, you may want to believe that your stuffed dog lives and breathes, but the reality is that it's just a toy.

Aligning your axioms with reality creates a stepping stone to help you better develop new theories. Even though you can't experience everything, your axioms can give you a more realistic view of the world.

A Marriage of Fate

On July 29, 1981, at the age of twenty, Diana, Princess of Wales, married Prince Charles. It became a global spectacle. Millions watched and fell in love with the new princess.[16] It seemed like a fairy-tale wedding.

However, there were problems from the start. Charles was in a relationship with Camilla Parker Bowles—a friendship, he claimed at first. Soon, though, it became obvious that Camilla and Charles were having an affair. Diana was the picture of grace, but, behind the scenes, she struggled with her husband's emotional unavailability and infidelities. She strove to maintain a perfect public image while dealing with these problems, but her attempts left her with anxiety and depression. She developed bulimia as a coping mechanism.[17]

Diana dove into humanitarian work. It was the escape she needed after a devastating few years stuck in a marriage that seemed like a trap.

Though the rest of her life would be fraught with turmoil, this one defining moment—her marriage to Prince Charles—was the most influential for the rest of her life. It was a decision made in her twenties that would forever impact how she behaved. There was a time when she discovered herself, and the early relationships she formed forever affected how she loved again.

Using the formative years of your twenties is a great analogy that expresses what it's like to be guided by passions and to learn more about yourself, but it's hardly the only time in your life that

---

16   Meilan Solly, "14 Fun Facts About Princess Diana's Wedding," *Smithsonian Magazine*, November 13, 2020, *https://www.smithsonianmag.com/history/14-fun-facts-about-princess-dianas-wedding-180976284/*. More than 750 million viewers worldwide tuned in.

17   Jane Mendle, "How Princess Diana Changed Lives by Discussing Her Mental Health," *TIME*, August 30, 2017, *https://time.com/4918729/princess-diana-mental-health-legacy/*. Diana described bulimia as a "symptom of what was going on in my marriage," and it continued for more than a decade.

this happens. In fact, we're always experiencing some revolution or another. We're always discovering more about ourselves and adapting our theories on life based on our experiences. In short, whether you're five years old or ninety-five, you're always changing, always evolving. There are always times when you shift your perspective based on new information. You never stop learning.

Chapter 4 Questions

- *How have your changing desires influenced the way you experience life?*

- *What beliefs or axioms have you developed to create a sense of personal meaning based on your thoughts, experiences, and emotions?*

- *What learning exercises do you use to help understand life when you question your beliefs?*

CHAPTER 5

How an Individual Makes Sense of Their Experiences— Developing a Theory of Individual Systematic Views

Our lives develop through the spectrum of experiences we accumulate. As our experiences grow, they work together to give us a bigger picture of our realities. As we make our way through our lives, our image of reality changes. The world seemed a whole lot bigger when you were three than when you are sixty.

Developing that personal reality is tough. Most of our experiences are difficult, and making sense of the lessons we learn is hardly easy. It takes a lot of perseverance to continually learn amid those complex situations.

But, as with most suffering, there is a purpose to it. The more we refine our views on reality, the better we understand our purpose in life. We get a better idea of who we are—our limitations, talents, and everything in between—and we align what we've learned about ourselves to develop an idea of what we can become. The experiences that failed either boosted your desire to continue or made you realize that you needed to take another path.

That's great. Defining your purpose is what most people strive for their entire lives. It's an answer to an age-old question: Why am I here?

But is that really the end of the conversation?

We've already established that your experiences define your reality, but does putting that definition in simple words change the way we understand life? Can understanding the purpose of life, as personal as it is, make life easier?

The "Normal Life"

Most people buy into a common stereotype: Life becomes more normal at age thirty. In your thirties, you're usually done with any education you want to complete, you might start having a family, and you're settling into your career. You've exited your crazy twenties and are embarking on (or are attempting to embark on) a life you can set your watch to.

Let's take a deep dive into this. You're commuting home after a long day at work. A common thought flickers across your mind: *I'm working way too much for way too little pay.* You wake up at the same time every morning, eat breakfast, say goodbye to your family, and head out the door. You get your coffee at the same coffee shop as you have for the last five years. You're pulled into the same meetings as always when you head into work. Your life has become routine.

Some will say slipping into the "boring phase" of life is where your ambition dies. You might think back to a time when you had a lot more fun and a lot fewer rules. And yet, there's satisfaction in doing the work.

As you pull up to your house, your brakes make a loud squeaking noise right before you put the car in park. *I need to make sure I get those brakes changed*, you think to yourself for the tenth time this week. You step out of your car and heave a familiar sigh. It's been a long day, and you're happy to finally be home.

As you make your way into your house, you're greeted with a home dirtier than you'd like it to be. You hear the squeal of a child, and you squint your eyes for a moment in frustration. Then, you see your daughter running toward you with a big grin on her face, and you smile right back.

In your thirties, your life becomes a lot less volatile. You've developed a life theory based on your experiences thus far, and you've started setting goals based on your values. Your thirties might be the time you reach out to a parent to get the wisdom you hadn't realized that they had before. Your life has changed a lot in just a few short years, and though the risks you take are much smaller, you still feel the need to reach out and discover more about yourself internally.

I'm not sure it needs to be said, but this is hardly a blueprint for everyone's life. And this time of internal reflection to develop a systematic view of life happens many times over a lifetime. For some—we might call them old souls—this time of self-reflection and development of a systematic view of life comes early. For others—we might say these people are children at heart—it takes a lot more personal experiences to settle on the basic system.

But just when you feel like you've figured life out, you make changes to your system to accommodate the changes in life. It may seem frustrating at first, but it's the path we need to take to discover our purpose. The small satisfactions, the changes in our perspectives, the experiences, and the trials all make us stronger. We find out why we're really here, something so personal it can't be found in a book. It becomes a goal we pursue for the rest of our lives.

Pursuing Life Goals

The ups and downs of multiple career changes and failed attempts at following in his family's footsteps led Vincent van Gogh to struggle in his early years. It was at the end of the 1800s that he attempted to build a life for himself. He shifted around frequently, working as an art dealer, a teacher, and a missionary.[18] Nothing seemed to really fit. He had some marginal success and satisfaction, but it didn't satisfy the itch of his true passion.

Van Gogh struggled with depression, anxiety, and psychosis. His violent temperament and intense personality often kept others at bay.[19] So, to ease the strain of his turbulent childhood and early adult years, he turned to art. He used it as a method to abate the loneliness he felt and the difficult relationships he had with his friends and family.

Van Gogh never received a formal art education in the traditional sense. Though he took a few classes here and there, he was primarily self-taught.[20] He spent his time looking through books and practicing. The paintbrush and the paint served as a mechanism to abate his inner

---

18   *Britannica*, "Vincent van Gogh," last updated December 8, 2025, *https://www.britannica.com/biography/Vincent-van-Gogh*. Van Gogh shifted through so many occupations that only the last decade of his life was dedicated to art, from 1880 to 1890.

19   Dietrich Blumer, "The Illness of Vincent van Gogh," *American Journal of Psychiatry* 159, no. 4 (2002): 519–26, *https://doi.org/10.1176/appi.ajp.159.4.519*. It's suggested that Van Gogh's mental illnesses were exacerbated by years of drinking and possible cranial damage.

20   "Van Gogh Essential Biography," *Van Gogh Experts, https://www.vangoghexperts.com/van-gogh-bio.html*. Van Gogh took a few classes at Académie Royale des Beaux-Arts in Brussels in 1880, but he quickly grew tired of formal study.

turmoil and emotional upheaval. Painting was constructive. Instead of lashing out at others, he could find the beauty in life.

It was during these turbulent times that Van Gogh painted some of his most iconic works. He used dynamic brush strokes and emotional intensity mixed with bold colors as the basis for his artistic expression. It was a revolutionary way to paint, and he completed over 2,100 artworks in his lifetime.

Though the end of Vincent van Gogh's life was tragic, coupled with the excruciating sting of never seeing success, he did create a legacy based on his passions. It would seem that, despite the experiences of pain and suffering, he had found his true purpose in life. One might even say that it was because of these intense experiences that he has become such an inspiration today. His vision of a starry night was an actual play on reality.[21]

The Correlation Response

Sticking to your passions is difficult. It takes a kind of courage that's years in the making. Vincent van Gogh only sold one painting in his life, but he inspired a style of painting that endured for another one hundred years and still retains its prevalence today. Though he might not have seen the fruits of his labors, he had made sense of his purpose.

For most of us, we already know what our purpose is, though we might not think of it that way. Each of us has developed a way to prove our theories based on our personal experiences. We've each developed a unique system that produces results. Over time, we begin to understand the correlation between our individual experiences and

---

21   "Starry Night: 10 Secrets of Vincent van Gogh Night Stars Painting," *Vincent van Gogh Gallery*, *https://www.vincentvangogh.org/starry-night.jsp#google_vignette*. Van Gogh painted *The Starry Night* from memory the next morning, so the image, though based on reality, is his interpretation.

what causes them. We start to see things through a unique lens that perceives things differently from anyone else's.

This ultimately breaks down into an optics concept. But, for the sake of simplicity, we can also think of this in terms of photography. Imagine you're sitting on a beach staring out over the ocean. You're looking at the sunset as it casts beautiful colors on the water. You just so happen to have taken your camera with you, and you hold it up to take a photograph. You take a picture of the sunset and look down at the camera to see the result. The image the camera took is but a depiction of the sunset created by the camera's lens and sensor. How closely the image and the sunset relate to each other is up to several factors: the lighting, the camera settings, and the perspective. For example, if you have a filter on your camera, it can alter the image, as you can easily understand.

In your own life, you have filters of your own. Someone with a flair for the emotional might see opera as the breakdown of human emotion and a kind of opulence expressing human artistry. On the other hand, someone with a more technical background may see the opera as a clever amalgamation of strategic note placements. Though the opera is presented in the same way to both people, it's enjoyed in completely different ways.

Identifying Life Correlations

Life is about finding the direct correlation between that object and that image. The thing is, our image of life will never be directly objective. Our thoughts, feelings, and perspectives are completely unique to us, so our view of anything is clouded.

That makes it seem like there is no way for us to find a perfect correlation between reality and our own perspectives. In a way, that's true. But that shouldn't stop us from finding our interpretation of the images we perceive.

In photography, we see the object as reality, such as with the Eiffel Tower. The image, formed by a lens, is our perception of the Eiffel Tower. In some cases, our image might look blurry. We may interpret the Eiffel Tower as having flaws based on the way the light hits it. We may see an incomplete view of the Eiffel Tower where our finger is covering the lens.

And we tend to do the same thing through our interpretation of logos, pathos, and ethos. In terms of our individual perception, the image is reality, and the lens is our interpretation of reality, which includes our thoughts, feelings, and perspectives. Though we all see the same reality, we interpret it through rational analysis, our emotional response, and our credibility and character, so we interpret things differently.

Let's return to the example of the image of the sunset. The images we see will always be slightly different from reality, but we can develop a system of interpreting our images consistently. For example, let's say the image you see of the sunset is depicted in sepia tones. The browns and beiges depicted aren't true to life. However, if you take an image of the beach that depicts the same level of sepia tones, you can start to see a direct correlation between those images. By analyzing the images you've created in your mind, you can start to see which filters you consistently put on your camera.

Now, not everyone will get it right the first time. In fact, most of us have distortions far greater than mere sepia tones in our personal understanding of life. We may see the world as a sequence of disjointed squiggles. Some of us may see reality in the photograph so completely differently from what the object presents that it takes some serious time to sort out the differences.

The best way to understand life better is to seek out the differences between objects and images to zero in on reality. We're seeking to find a theory on life that highly correlates with established facts. That's the only way that we can comprehensively predict how we should behave in different situations. This takes a lot of work. We have to start analyzing which objective observations or data points are verifiable and consistent. Next, we need to actually incorporate them into our personal theories. We're building a firm foundation on the facts. Sometimes, that means looking back and forth from reality to our perspective through perceptions multiple times before we find a systematic approach to living our lives.

The good news is that we don't have to do this bit on our own. We can use the experiences and lives of others as the starting point for the way that we understand the world. Our experiments, observations, and interactions with the world are a way for us to gather facts and create hypotheses that will ultimately refine our theories. But, if we can use others as a guiding point to help us better understand ourselves and the world around us, we can get there even faster.

One of the most beneficial tactics to help us refine our theories is seeing something from another perspective. We already know that we can never see life in a truly objective manner. We can start to see the faults in our own theories when we see the discoloration in someone else's images. It's like giving ourselves a new set of eyes with a brand-new way of thinking about things.

Once we've established a theory that provides us with a systematic view of life, we can start incorporating new experiences into our axioms based on whether or not they fit our understanding of life. Is being late to a dinner party the result of your friend's consistent lack of punctuality, or is it because something truly terrible has happened? When you start to see things in patterns (occasionally borrowing someone else's perspectives), you can start making much more rational decisions.

Refocusing on Purpose

That's a lot to swallow, so let's take a step back. As humans, we are drawn to developing an understanding of our surroundings by understanding purpose. We want solutions to problems that have reasonably limited scopes. Just as in the example from the previous chapter, you don't want to spend the rest of your life thinking of reasons why the bus was late. Eventually, you need to come up with a reasonable solution that matches previous experiences and serves your purpose.

As you experience more, you can better understand which explanations work best for you, and this leads to consistent theories.

It's a fact of life that some things are worth more thought than others. People have spent their entire lifetime trying to come up with a universal purpose for life, while others will stand in front of a hardware store for hours trying to find the right paint for their living rooms. For the latter, the distinct lack of purpose is evident in most cases, if not all. Yes, you want your living room to look nice, but there are much better things to focus on, unless the shade of the wall may affect your autistic son's development, for instance. Otherwise, there's more to life than simple shades of white.

It's definitely a challenge at first to aim for a purpose that will satisfy your time. Some inconsequential experiences may keep us from identifying what's really important. It's easier to get stuck in the small details.

Elevating Your Outcomes

The real elevation of our experiences comes from focusing on positive outcomes. Let's do a bit of a thought experiment. Think back to the last time you made a major decision. Did something bad happen to you, or was it the result of your own actions? Do you think back on that experience with fondness, or do you feel a bit jaded by the results?

Many of us tend to think about the experiences in our pasts—especially those with some of the worst outcomes—as things to forget. Dwelling on the past isn't always fun, and it can result in embarrassment and frustration as well as joy or happiness. For those experiences that caused a bit of emotional, mental, or physical damage, it's easy to think negatively of the decisions made.

But what if, instead, we focus on the positive aspects of each of those decisions? Perhaps the decision you made wasn't perfect, but surely it had some redeeming elements. You chose it for a reason, right?

To truly elevate how we experience the world, we need to make an adjustment to the way that we assess our experiences. Ultimately, it comes down to realigning the outcomes to match the assumptions. Your past experience taught you something, even if that is just what not to do next time.

Let's think about it this way: Imagine you're training to run a marathon. It's the first time you've ever trained, and you start roughly eight weeks before the race. You've competed in the swimming portion of a team triathlon in the past, so preparing for a race is an experience you've already had. You start by running five miles a day for a week. By the end of week two, you're running seven miles a day. By the end of week five, you've reached fifteen miles per day. By the time the race comes around, you've been running twenty-two miles per day. It's not quite where you want to be, but you have done a fair bit of work to be where you are today.

You've been training in Eastern California, but you've always wanted to run a marathon in Boston. After your flight, you feel ready to get started. But it's not quite what you expected. Boston is much more humid than you're used to, and you feel your tongue start to stick to the back of your throat.

Still, it's something you've always wanted to do, so you suck it up and start running.

After mile five, you're starting to feel a little more out of breath than you thought you would be. Even on the hottest days during your preparation, you never sweated like this. Every time you pass a water station, you take two cups, one to toss down your gullet as fast as you can and the other to dump all over your face.

At mile ten, you start to think about quitting. It sounds so much nicer to sit on the sidelines and cheer on some of the other people, who are used to Boston weather. But you shake your head and remind yourself that you've been preparing for this for two months.

At mile fifteen, you have to start directing people to go around you. Your sweat is leaving a trail, and you've become slightly alarmed at just how much water you've been losing. You'd be in serious trouble if your long line of sweat was the reason for someone slipping and falling.

At mile seventeen and a half, you have to call it quits. You've worked as hard as you possibly can, and you're barely dragging yourself to the sidelines. It's a moment of deep reflection for you. While this marathon was something that you've always wanted to do, you definitely weren't prepared.

When you get home, you feel lousy about not completing the race. But it's taught you a valuable lesson. You're determined to finish the Boston Marathon. You just need to adjust your assumptions to fit your outcome. The next time you start to train, you start four months before the next Boston Marathon. You travel to locations with higher elevations and others with humid conditions. Instead of reaching a "good enough" distance of twenty-two miles at the end of your training, you elect to run a little bit further, ultimately completing just over thirty miles at your longest stretch. When the next Boston Marathon comes around, you're finally ready.

Realigning goals isn't easy. It requires us to take a good look at our purposes and find solutions that will match them. Those experiences you had with failure may have seemed like roadblocks, but instead, they were the moments that caused you to refocus.

The Goal of Self-Consistency

As much as we would like to think they're black and white, human interactions ultimately have multiple facets. It seems almost contradictory to say that we can develop a self-consistent system and yet adapt to the different situations life puts us in. We can't go our entire lifetimes believing that a single limited scope is enough to create a system that will work for the rest of our lives.

We can, however, become much better at adapting our purposes based on different contexts.

The goal is to break down your purpose into adaptable goals and principles. Find priorities that are naturally flexible and relevant. For example, prioritizing helping your community can take a variety of forms and can change based on different contexts. It can also help you narrow down exactly what you want to do in life. Think of it like a thermometer that gauges the temperature of your purpose. When you feel more connected, more open, and more free, you know you're hitting the right temperature in your gauge.

But don't be afraid to experiment. We make the best decisions when we have a diverse base of experiences to gauge our final decisions. I like to think of it this way: You should take in everything the world has to offer and then pick and choose which works best for you. You're the master of your own life, so pick options that will help you reach your goals and honor your priorities.

Where most people find problems is in their flagrant rigidity. It's easy to remain fixed on a belief at any point in life. In fact, you may have heard the excuse, "I was just born this way." That's not the resounding war cry that most people think it is. Actually, it's a sign of severe limitation.

An adaptable growth mindset welcomes challenges. Those with rigid thoughts and stagnant emotional responses don't bend under pressure; they snap. So, to maintain a self-consistent system that will stand the test of time, create a sense of emotional intelligence and resilience that will sustain even the toughest criticism.

The Advent of Courage

Perhaps known best for his leadership during World War II, Winston Churchill has always seemed a working character. To many, his name is synonymous with resilience, leadership, and indomitable spirit. And yet, that wasn't always so.

Churchill found a consistent lifestyle after marrying his wife in 1908, when he was thirty-three.[22] He was deeply involved in his children's upbringing, which instilled in him a strong sense of duty. He ultimately lived to serve his children, and having a strong sense of justice and freedom was paramount to becoming the right moral

---

22   *Encyclopedia.com*, "Churchill, Clementine (1885–1977)," last
     updated December 1, 2025, *https://www.encyclopedia.com/women/
     encyclopedias-almanacs-transcripts-and-maps/churchill-clementine-1885-1977.*

figurehead.[23] His new sense of courage came in response to his desire to better serve his family and country.

Churchill developed his self-consistent system of values that guided his personal and professional life after there was some stability in the home. Miraculously, it was a self-consistent system that lasted well into his sixties, when World War II broke out. Having a system that stands up under immense pressure is the mark of a man who developed his system early and practices it daily.[24]

You may see a bit of yourself in that story. After all, many people develop certain values and self-consistent systems early in life. For others, it takes some time. There's never a right or wrong time to develop your system. We all experience life in different phases and at different times. Those systems often change over time as well. So, it wouldn't be surprising if you found yourself developing a whole new system even in your nineties. It takes courage to change value systems and create a system that works for you, but it's a courage anyone can develop if they have the will to try.

---

23   Justin D. Lyons, "Winston Churchill's Moral and Philosophical Guides," *The Churchill Project, Hillsdale College,* November 8, 2021, *https://winstonchurchill.hillsdale.edu/moral-philosophy/.* Churchill studied Plato, Aristotle, Gibbon, and others to better analyze his philosophical basis.

24   Cole Feix, "Churchill's Character: A Rigid Daily Schedule," *The Churchill Project, Hillsdale College,* February 6, 2019, *https://winstonchurchill.hillsdale.edu/churchill-character-daily-schedule/.* Churchill took self-discipline very seriously, extensively reading, writing, and working to understand his political failures and reflect on his leadership.

Chapter 5 Questions

- *How have you seen your viewpoint of the world change when you redefine your purpose?*

- *How have you developed courage and a sense of resilience by identifying what you need to change about your lifestyle and seeing things from a new perspective?*

- *How have you become more flexible over time, and how have you become more rigid?*

CHAPTER 6
Why Self-Consistency Is Important, and Why It Is Essentially Not Possible

Self-consistency is the cornerstone of our perception. We see the world through a lens that is unique to each of us. The choices we make—and the people we will become—are all because of this self-consistent system.

But, in reality, there are always perceptions we will never know. Though we may wish to understand what life is like for those who live in sub-Saharan Africa, 98 percent of the world won't have that chance. At the same time, any conclusions that we draw from our perceptions will always have a modicum of incorrectness. If we can't see the full spectrum of the world, how can we expect to create a flawless system that requires an omniscient gaze?

Now, this may seem like a difficult question, especially after we spent an entire chapter talking about the importance of creating a self-consistent system. It seems counterintuitive.

Well, that's something we have to get used to as humans.

It's important to note that a self-consistent system is, by definition, limited. The scope of the world has an infinite number of perceptions. Who we are in this moment in history can never be replicated, not even by someone living in this exact place and making similar decisions twenty years from now. And that's something that we just have to accept.

The Center Perspective

At the ages of forty to forty-nine, a unique thing happens. Your life is suddenly half over, and there's a lot to consider. You've just finished the decade that cemented a more regulated lifestyle. There were lots of ups and downs, successes and failures, during that time. It's at this time that most people start thinking back and considering what might have been.

Let's take a look at that.

You pull into the driveway of the home you've owned for the last decade or so. Your nearly grown kids are inside, and it's eerily quiet. You walk through the front door and kiss your spouse. You haven't seen them all day, but it's still a feeling of routine that you barely think about anymore. You poke your head into the room of your oldest child, and you see them playing a video game with their headphones on. You call out to them that you're home, but they don't seem to hear you. That's nothing new, either.

As you walk up the stairs of your home, you notice your feet take the same path they always do. You finally make it to your bedroom door, and you push it open to hear the familiar squeak. You begin changing into pajamas and slippers, just something to remove yourself from the workday.

You hear yourself sigh as you walk over to the window and look out at the car that you drive every day. It's a boring shade of gray, much like the suit that you put on daily. *I'm living a comfortable lifestyle right now,* you think to yourself. When you were struggling in your twenties, you always wished that you would make it to this point. True, there have always been bills to worry about, but there's a much greater sense of relief in settling into a nice routine. Just like your parents before you, you can pay for the things you need and even afford a little extra.

But what does it all mean?

As you look slightly to the left of where your boring gray car is parked in the driveway, you see that your neighbor has just purchased a new Corvette. It's shiny and red. You know it's a midlife-crisis purchase, but you can't help but be a little bit jealous. You notice that your neighbor's wife is out playing with her kids. They're jumping through a sprinkler and laughing. It looks like the perfect picture.

Suddenly, you start to question what you've been doing this whole time. You've been struggling to perfect your life. You realize that you were so incredibly absorbed in what should be that you failed to comprehend what is.

You take a deep breath and puff out your chest as you head down the stairs to your eldest child's room. You tap on their shoulder, and they barely turn to look at you. "What is it?" they ask distractedly.

"Take off those headphones," you say. "We're going mini golfing."

Many of us have an idea of perfection in our lives. We're so obsessed with completing a goal, with achieving perfection, that sometimes, we don't stop to realize there is some adaptability in our lives. It's like we've become so hardened to believe that we can only live life in a way that we've perceived is the most valuable that we fail to realize our self-consistency isn't entirely possible. In reality, there is no perfection. Because we don't have all the answers and we can't see everything perfectly, we will never have a self-consistent system that works for everything.

Fortunately or unfortunately—however you want to see it—we come across this dilemma multiple times in our lifetimes. A nine-year-old may be just as insistent on mastering the violin as someone aged ninety. And in both cases, where they're stuck in their ways, it's impossible to achieve true perfection.

For many, the most difficult part of that dilemma is realizing that, sometimes, they may just have to let go. And it can be most rewarding to do so.

The Mandela Effect

Letting go requires a tireless process of true lifestyle evaluations. We can believe something as children, but the more we experience in the world, the more we adapt based on new evidence. This was never more evident than in the apartheid-ridden South Africa in the early twentieth century.

Nelson Mandela, born Rolihlahla, started life very familiar with government. His father served as the local chief and counselor to the royal family, bringing with it a sense of responsibility and leadership. Mandela's lineage was special, with its key feature being familiarity with the governance structure of the Thembu, the royal family.[25] At a young age, he was taught to understand justice and to develop a deep understanding of African leadership principles.

He studied relentlessly in school. His rigorous academic curriculum included the development of his leadership skills and gave him a bird's-eye view of the racial inequities that were pervasive in

---

25   "Biography of Nelson Mandela," *Nelson Mandela Foundation, https://www.nelson-mandela.org/biography.*

South African society. It was then that he participated in the active boycotting of university policies. His involvement led to his expulsion.

His deep desire for involvement in rectifying injustices led him to the African National Congress. It was the first step in prompting him to open up the first Black law firm in Johannesburg in 1952 to reduce the effects of the apartheid laws.[26] Subsequent upsets and the Sharpeville massacre[27] led to increased resistance against apartheid, and Mandela was forced underground before his ultimate arrest a year later.

The next twenty-seven years of imprisonment forced a new Nelson Mandela to arise.

He'd always wanted to create a new South Africa. Instead of a fight against power, he became enamored with the idea of a nonracial, democratic South Africa, free from the petty fights and constant warring. The years of violence had ultimately achieved nothing. Perhaps, instead, a future that consisted of negotiation and diplomacy over violence would eventually prevail. With the right people and a lowered intonation of the voice, he could facilitate a peaceful transition to democracy.

A mere four years after he was released from prison, the dismantling of apartheid laws allowed his inauguration as the first Black president of South Africa.[28]

---

26 Nelson Mandela, *Long Walk to Freedom: The Autobiography of Nelson Mandela* (Little, Brown and Company, 1994). The law firm was seen as "the firm of first choice and last resort for Africans."

27 Arianna Lissoni, "*Sharpeville: An Apartheid Massacre and Its Consequences,*" *African Affairs* 111, no. 445 (2012): 686–87, *https://doi.org/10.1093/afraf/ads045.* On March 21, 1960, police opened fire on demonstrators, leading to the death of sixty-seven people and the wounding of hundreds more.

28 "Life and Times of Mandela," South African Government, *https://www.gov.za/ mandela100/biography.* Rolihlahla Nelson Dalibhunga Mandela was inaugurated as the South African president on May 10, 1994.

But perhaps the most important achievement of his life, which he would mention as a vitally important consequence of his imprisonment, was his ability to forgive. He became an ambassador for peace that extended far beyond South Africa. He embraced his role as a global ambassador, constantly advocating for social justice and human rights worldwide. It was a concerted effort to ensure his message of forgiveness and unity was given throughout the world.

It's interesting that forgiveness, something so pronounced in the rest of his life, wasn't considered as his primary platform when he started his law firm. It took backbreaking hardship to realize the path he had to fulfill: the path of unity and peace.

You may see yourself somewhat in Nelson Mandela's story. In the twenty-seven years he was in prison, he had the opportunity to look back on his life and question the way he had grown up. For all of us, at some point in our lives, we realize that the way we have been living isn't entirely consistent with what we want to be. Though Mandela spent the rest of his life working toward creating a world that valued peace, it was something he would never see achieved. Most of us have a modicum of that reality in our lives. We'll constantly work toward a goal but never see a perfect completion. At the end of the day, we all have a long way to go.

Aiming for Self-Consistency

For most of us, that idea is rather sad. After all, we are hoping to become as perfect as possible. We want to leave this world knowing we have accomplished the goals we set forth to complete. But the reality is that it's not that easy.

Let's say that we're looking back at the history of the Roman Empire. Rome's peak was roughly two thousand years ago, so we should have a bird's-eye view of what went right and what went wrong. We can see the devastating effects of emperors such as Nero and the wisdom of those such as Marcus Aurelius. Of course, the only way we can learn about these people is to trust the words of those who have their own perspectives. The Roman writers saw the Germanic battles as successes, but we don't really know what the Germanic people thought. There's a glaring hole in the commentary that would give us a more rounded view.

Now, let's assume that you want to understand more about the Roman Empire. You can read as many books as possible on the subject, taking into account as many perspectives as available, both from the winning and the losing side. With all your research, you might get a fragment of the true story of the rise and fall of the Roman Empire.

The truth is, some factors are outside of your control. Some things are truly impossible to figure out after two thousand years, and our theories will always have inconsistencies. The image by which we see the world will never be 100 percent accurate.

You can drive yourself crazy trying to be as precise as possible. PhD students may do that if their research is about the Roman Empire. Others don't have to do that. Even if we take an example as simple as understanding why a friend was late for her lunch meeting, you have limitations in your knowledge. Your friend might tell you the reason why they were late, but you'll still never understand the true story, as there are so many details left unsaid.

So, the best option is to simply stop.

Understanding the Limitations of Self-Consistency

We need to be realistic in our lives. There is a limit to self-consistency. Whatever theory we develop and the system we put in place will inevitably be broken when we learn more and expand our boundaries.

There is only so much manipulation that we can do in our daily lives of things outside of our scope of understanding. While we might want to know what the weather was like in Rome on October 7, 1645, it's simply outside of our reach. We can extrapolate what the weather might have been like at that point, but we'll never know for sure. And if we spend too much time focusing on those minute details, we'll drive ourselves crazy.

But let's use an example that's a little closer to home. Let's say you were given a lock to protect some of your valuables when you were young—a journal, your piggy bank, or really anything else that you considered valuable. Twenty years later, you return to your childhood room and discover that small lock you were given many years ago. You remember that you had a special way to remember that lock combination. Maybe it was the birthday of someone close to you or an anagram inspired by a favorite childhood book. As you stare at the lock, you can't seem to remember exactly what the digits are.

So, what failed here? At least at one point, you had a self-consistent system that helped you remember the combination of that lock. You tried it, tested it, and then applied it to your daily life. But, twenty years later, it's not coming to mind. You could spend years trying to figure out exactly why you forgot the combination. There are nearly an infinite number of explanations for your lapse in memory. But the simple fact is that you forgot. And you forgot because your self-consistent system has evolved.

As humans, we are simply not infallible. Our systems inevitably fail. We already know that, given that our scope is always limited, there will always be problems with what we establish in our self-consistent systems.

But where do we go from here?

Centering on Happiness

The key to happiness is to find a point at which we can let go with satisfaction. You need to be satisfied knowing that your curiosities never will be. After all, at some point, we have to let go when leaving life, whether we like it or not.

You may be thinking, *That's easier said than done.* Just letting go of curiosity isn't in our human nature. Each library is a testament to our never-ending search for knowledge. And the concept may sound strange coming from a STEM-educated person who has spent his whole life dedicated to finding answers to questions. But here is what I suggest: You have to put the work in.

Think about this for a moment and see if it resonates with you: Many of us spend much of our lives wishing we weren't in one situation or another. Perhaps we wish we had a better job or more money or lived in a better place. We spend so much time wishing we were somewhere else that we fail to realize the beauty of where we are. It has become so easy to wish we were somewhere else that thinking otherwise is a chore.

So, we need to develop a system that works well enough for us and leave it at that.

As much as we would like to think otherwise, we can't find a self-consistent system that will work 100 percent of the time. We need to be more adaptable, more willing to open up to different possibilities. It's a conflict that we all must battle daily. In essence, it's the fight to choose happiness over mastering every perspective.

Operating Solo

Our never-ending striving for happiness ultimately comes with a unique bonus. The less time we spend focusing on trying to solve every question or finding a solution that will fit every scope, the more we ultimately achieve. In essence, we're relying on ourselves to find solutions that will fit limited scopes. The more we can accept, the faster we can move.

Nelson Mandela's work as the president of South Africa was very limited. He only retained the position for a few years, as he found his time was better spent on finding a solution to unity. His quest was an impossible task. There was no way he would be able to find a solution to all racial inequality or violence in the world, especially not working alone. He was extremely limited in what he could do himself. Though he had a very extensive background working with a variety of people, he couldn't change the perspectives of everyone.

Still, this limitation wasn't enough to stop him. His last twenty or so years of life led to the Nelson Mandela Foundation in 1999, HIV/AIDS advocacy and support, international diplomacy, and many more developments. If he couldn't be the one to change the world with his own hands, he provided the voice for the people who could. He found a new way to achieve more when he let go of the way that he saw his limitations. He achieved far more by accepting his limitations and not relying on others to make important decisions for him than he ever could have done waiting for systems to right themselves.

Ultimately, the journey is up to you. We can read as many books as possible about how to be happy, but the only way to be happy is to put in your own work. You can't rely on anyone else to make you happy. The journey is there for your own benefit. It takes work, but it's worth it.

An Eye on the Prize

The application of finding your own happiness is actually much simpler than many would lead us to believe. It all starts with recognizing the limitations of our views. So, before we completely revamp our self-consistent systems, recognize that you don't have all of the answers, and that is perfectly OK. We are designed to form opinions based on the best knowledge available to us.

Next, we need to recognize the inconsistencies in our priorities. What we want in life constantly changes. Who we were ten days ago isn't who we are now. Maybe natural disasters have changed our desire to buy that new vehicle. Maybe the death of a loved one has made us insist on spending more time with the people we love. Whatever the case, we need to accept that those priorities change. We can't have an unshakable system. There are always things subject to change. We evolve as our self-consistent system does.

So, if we can't change our perspectives, it only stands to reason that we need to better understand how we can manipulate the world around us. No, we can't stop others around us from acting on life, but we can change our mindsets to better accommodate the shifting tides of life.

For millennia, philosophers have insisted that we are not as in control of our lives as we think we are. Buddhist philosophers from eons ago have stated that our control is limited, and the only way to happiness is to let go. Stoic philosophers have insisted that we cannot control external events, only our responses to them. Modern philosophers have suggested that true freedom comes from recognizing and transcending the belief that we are in control of our lives. Regardless of the time period, the greatest minds have come to a simple conclusion: Our search for happiness shouldn't rely on believing that we can control how it is brought to pass.

I want to take this universal striving for happiness a little further. What we can create, who we strive to be, is a result of our deliberate efforts to be happy. A billionaire can be distraught because he lost the bid on a unique art piece at the auction, and a homeless man living off almost nothing can derive joy from watching a dog chase its tail.

True happiness is an understanding of contentment and the act of letting go of perfection.

A Stoic Mind

Marcus Aurelius struggled with the pressure of responsibilities that came with ruling the Roman Empire. He sought to be deeply in control of who he was, a stark contrast to his predecessor. He focused his life on maintaining a sense of virtue and duty, fulfilling a life of integrity, and maintaining a steady environment for his people.

But he struggled deeply throughout his life to maintain a high level of stoic resolve.

Marcus experienced the deaths of several children, and he struggled with profound grief. His surviving son, Commodus, didn't embody the same beliefs that Marcus Aurelius did, and he fought with himself regarding the succession. He wanted to pass on his personal beliefs—a philosophy he wrote down in a book that still survives—to the next generation, perhaps influencing external events and personal choices that might occur.

And yet, as an adherent of Stoic philosophy, he knew that, at some point, he had to let go. Life was unpredictable, and he couldn't always rule over his people. Roman succession didn't require a blood relative, but the threat of Germanic tribes encroaching on Rome's territories and his inability to find a suitable successor meant that the crown passed naturally to Commodus. He knew that his son's erratic behavior and self-indulgence would lead to problems within the empire, but there was no other choice. His final gift to the world would be the writings he passed down.

Many throughout their lives often feel the same lack of control. It's most often depicted as a midlife crisis, but that's hardly the only time frame to feel subjugated to perfection. We often want to find a solution to problems, and sometimes, that means clinging to unpredictability for too long.

Knowing that we'll have problems and will struggle with perfection is just a part of life. The greatest way to overcome unpredictability is to just let go. Though we make the decisions in our lives, we're ultimately not holding the reins to our fate.

Chapter 6 Questions

- *What are some of the most difficult problems you've had to overcome by letting go?*

- *What aspects of perfectionism are you still clinging to, and what steps can you take to let go of those feelings?*

- *When was the last time you struggled to ensure you were truly happy?*

CHAPTER 7

How to Resolve the Conflict of Self-Consistency

It's easy to say we need to let go as a resolution to the conflict in our resiliency. But the question remains: How do we choose to adapt to our circumstances?

The real question is, when do we let go? If we let go too early, we won't achieve anything, while letting go too late means we won't achieve in other aspects of our lives. Letting go isn't pessimistic. In the long run, it's the best way to achieve as much as possible in life. The best way to maximize achievements is to set the right criteria.

Our internal criteria play a big part in our conflicts. They help us understand the adaptability of our natures, and they provide a big step forward in the right direction. We can better understand how to resolve the paradoxes that come from the lack of self-consistency by analyzing why we think the way we do.

The first step—really the first step we've seen in this book as well—is to develop theories. That process never really stops. We need to have a bird's-eye view of the scope. Every time we see a flaw in our logic, we upgrade our theory to a new level that will solve that level of consistency. It's a series of baby steps that incorporates abstract thought.

If you've gone through this process—and many of us have, many times—you might notice a troubling theme. When you're expanding your scope, it doesn't always help you solve your problems. You're essentially laying the groundwork for a future consistent system, and that means there will be hiccups. There will be problems you can't foresee, and a solution outside of your scope is difficult to put in words.

But focusing on the problem helps you overcome the uncomfortable trial-and-error phase. Remember, even the brightest minds in history have come up with truly terrible conclusions on their way to solving lifelong dilemmas.

A Sense of Reinvestment

When we enter our fifties, there's a part of us that accepts a new phase in our lives. Most of the children are already out of the house, and you've already spent thirty-something years in the workforce. It's time you had a break, right?

Picture this.

You're sitting in an Adirondack chair, looking over at the mist covering the mountains in the distance. It's a gift you've given yourself and those around you whom you love: You've finally taken everyone on the vacation you've been dreaming about for many years. Your kids are already grown up, but they've come back for the cabin trip. It was a lot for them just to get out of work, but you made sure to drill into their heads that this was a priority.

You take a deep sigh as you gaze over at the green grass, newly illuminated by the rising sun. It seems like work has been your number one priority for so long that you don't know your family as well as you wish you did. This trip is a step in the right direction.

It's the perfect relaxing environment to get your mind off the fact that you're thinking about retirement. Your priorities are going to change, and it's scary. You're finally seeing resolutions when it just seemed like an endless push into the future before. Yes, you'll have more time, but your life will be completely different.

You hear the door squeak, and your daughter comes out to sit next to you. "Amanda told me she's getting married in a few months," she tells you. Amanda had been her best friend in elementary school, and they had remained best friends through high school, but they had lost touch when each of them went off to college. "She asked me to be a bridesmaid. Do you think I should go?" It's been a long time since she's confided in you.

You sit back, put your hand on her knee, and heave a deep sigh again. "Yes, you should go. These kinds of moments never come again."

"Yeah," she says, "maybe I should." She stands up and walks back into the house.

The advice you gave sounds strange to you. You've spent so much of your life trying to provide for your family, but you realize you've missed some key moments. You wish that you had been there when your best friend got married. Maybe you're finally growing wiser.

Most of us experience something like this at multiple stages in our lives. We've stayed in the same system for so long that habits form. It's a struggle to accept changes, even if we see them coming. Many times, those struggles seem paradoxical, unsolvable. And just when it seems that you've found a resolution, the scope changes.

Unfortunately, life doesn't always give us the time to reflect on true solutions to every problem. The only way we can continue to move forward is to expand that scope a little at a time.

Each Virtuous Step Forward

Each of us has striven to better ourselves in some way. Growth only comes as a result of sacrifice and struggle. It's a concept Benjamin Franklin knew all too well. He spent his entire life focusing on education and self-improvement, largely influenced by his Puritan upbringing and the Enlightenment ideals of reason and progress.[29]

In 1726, at the age of twenty, Benjamin Franklin began a period of personal development and self-discipline. He began to truly reflect on his life and plan for the future. It was merely months later that he developed a plan in the form of a list of thirteen virtues.

The thirteen virtues were inspired by the desire to achieve moral perfection and to improve his character, the progress of which was illustrated in his autobiography. Each virtue was foundational to developing good character in daily living, which meant that each virtue could be tested daily to improve his behavioral and societal interactions. Temperance focused on moderation and consumption and promoting health. Silence encouraged active listening and less unnecessary chatter. Order emphasized organization and time management. Resolution stressed the importance of setting goals and following through. Frugality advocated for financial management and resource conservation. Industry promoted productivity. Sincerity encouraged honesty and integrity in thoughts and communication. Justice focused on fairness and fulfilling obligations. Moderation advocated for a restraint in reactions and behavior, a balance in all things. Cleanliness emphasized personal hygiene and

---

29　Thomas S. Kidd, "A 'Thorough Deist?' The Religious Life of Benjamin Franklin," *Age of Revolutions*, June 5, 2017, *https://ageofrevolutions.com/2017/06/05/a-thorough-deist-the-religious-life-of-benjamin-franklin/*. Benjamin Franklin pioneered many publishing, scientific, and diplomatic movements in the United States and gained much of his dedication to learning from his criticism of Puritan sermons.

environmental tidiness. Tranquility encouraged maintaining calmness and composure. Chastity promoted responsible and respectful behavior in intimate contact. Finally, humility encouraged modesty and the recognition of one's limitations.[30]

This long list is daunting when taken at face value. So, to make the progress a little less unwieldy, Franklin devised a plan to focus on one virtue per week. He maintained a methodical approach to marking his successes and failures, concentrating on developing each virtue while maintaining an awareness of his overall progress.

As with any goal, there were periods of setbacks. These he framed as opportunities for learning. Benjamin Franklin was focused primarily on lifelong improvements and progress, which meant adapting to new challenges and opportunities.

What perhaps began as a moral experiment on the improvement of human character soon became a lifelong guide to reason and self-reflection. Benjamin Franklin is perhaps best known for his other achievements: the use of a key and a kite to find electricity, his diplomacy overseas, and his autobiography. And yet, what his life truly showed was the power of the mind to overcome great internal and external battles.

Benjamin Franklin's innate desire to achieve more through discipline is one of the greatest examples of pursuing through paradoxical self-consistency conflicts. It's not a difficult process. It comes down to maintaining a standard of achievement through small steps.

---

30 Benjamin Franklin, *The Autobiography of Benjamin Franklin*, ed. Leonard W. Labaree et al. (Yale University Press, 2003). Franklin's thirteen values served as the basis for maintaining discipline.

A Deep Dive into Solving Inconsistencies

Our lives are never-ending cycles of dynamic changes in beliefs and experiences. We constantly experience flaws in our logic and new data. So, it should come as no surprise that our self-consistent systems encounter obstacles on a daily basis. In fact, solutions that may have worked for some problems may not work for seemingly identical issues.

So, how can we reconcile the challenge of maintaining self-consistency when everything around us, including ourselves, is constantly changing?

You may have heard the famous saying, "You can't step in the same river twice." It's an ancient idea, dating back to ancient Greece, that shows both we and our environments are in constant flux. Often, the information we think we know changes over time. Sometimes, what we see as our past experiences and our current realities seem to be at odds. During these times, we often feel tension and confusion. It's a struggle to reconcile our desire for stability with the inevitability of change.

At this point, though, it's important to take a step back. This is just another dimension in our understanding.

Finding Dimensions

Imagine, if you will, you're taking a trip to Las Vegas. You're there to see the sights and see some of the local art. While you are traversing the Strip, you stop when you see a spray-paint artist creating an image from scratch. He starts with a blank white canvas and then paints it black. Next, he takes a piece of cardboard with a circle cut out and lays it on top of the black surface. He first spray-paints yellow, then orange, then green in sweeping motions, finally pulling up the cardboard after the entirety of the circle is filled. He repeats this with another circle, this time spray-painting with pinks, purples, and blues. Next, he places a large circular pan over each of the circles and outlines rings. Finally, he takes an old paintbrush, dips it in white paint, and flicks it over the canvas. What once was just a few shapes on the canvas has now become an outer space scene.

When you first approached his table, it was hard to see what he was going to create based on the tools he had. What's more, you couldn't see the final image based on the individual steps it took to get to the final scene. We experience life in a series of dimensions. It's usually not clear what the final image will be.

Unlike this example, we often have inconsistencies when drawing on our own canvases. To us, the image seems incomplete. And, as is often the case, there usually are things that we are missing.

But what if we widened the scope?

If we only looked at one of the circles painted in a series of colors that don't look like they correlate with one another, we're not getting the full picture. But, if we zoom out to the full size of the canvas, we get a broader scope of what's really there. Instead of merely a circle, we can now see a planet. Instead of unrelated colors, we start to see the shadows cast in space. Even something so static as a spray-painted canvas can come to life and tell stories if we zoom out and analyze the scope in greater detail.

We can see something similar in our own lives, if not quite that simple.

The key is looking for those added dimensions. If our scope is too narrow, it's often difficult to find ways to add perspective. Seeking to understand by pushing your own comfort zone, your own boundaries, aids in helping you identify the extra dimensions you might have missed.

Setting the Criteria

Setting criteria to reduce the inconsistencies in your self-consistent system helps you clarify your judgments and decisions. You can better understand how you evaluate systems and what leads to those judgments. Establishing a set of criteria in your current scope of understanding gives you the freedom to measure and evaluate new information, ideas, or actions.

Let's consider, for example, Benjamin Franklin's pairing of silence and justice as virtues. When you see an injustice committed, what is the first decision that comes to mind? Silence is a virtue, but how can you remain quiet when trying to fight for justice? The answer might be listening to different opinions and understanding each party's arguments while maintaining your fight for justice.

Self-consistency means finding a way to incorporate all your beliefs. Do the elements of your criteria align with the new input?

First, consider the key components and relationships of new challenges with what you already know. When you test your hypothesis, do you find inconsistencies? Reflect on the factors that influence your judgments. You might find that certain sources of information and perspectives shape your understanding of your system more than others. Are your criteria based on logical reasoning or your personal values?

Next, start to ask yourself whether your criteria are comprehensive and inclusive. We know that there are different dimensions to our scopes. You might find that refining and adjusting your criteria can give you a better foundation when you evaluate your system's consistency. When you see potential inconsistencies and conflicts—paradoxes—you can adjust your scope to include the new elements.

The clearer you are when defining your criteria, the faster you can adapt to different situations. If most of your criteria are based on logical reasoning, evaluate if your reasoning is coherent or flawed. If your criteria include personal beliefs, periodically make sure that your value systems line up with them.

Prioritizing the Experiences That Work for You

Of course, we want to be as logical as possible when assuming a self-consistent system. That logical response necessitates an objective viewpoint, but that's essentially not possible. Because we all have different perspectives, we highlight, intentionally or unintentionally, the most important parts of each of our experiences.

Let's return to the example in chapter 5: the image versus the object. When we look at the image taken from our personal perspectives, we see a somewhat distorted image of the object. In our image, we've highlighted different portions of that sunset than reality shows. Some may see the image of the sunset in a range of rainbow colors, while others see it in black and white. Each person has decided, consciously or unconsciously, which part of the image to emphasize. As a result, their perspectives have forever colored that image and distorted reality. This is also true for geometric optics, as the lens, or the optical system, can never collect all the light rays reflected from the object to form an ideally perfect image.

That may sound like a terrible waste. We've seen a perfectly good sunset, so why can't all of us see it in the same way?

There's something profound in seeing things from a different perspective. We develop judgments of the world based on those differing perspectives. Perhaps the person with an overly colored sunset image sees the world as a much more emotional place than the person with the black-and-white image. At the same time, the latter person may see nuances beyond emotional responses that shed light on logical systems.

Whatever the case, it's clear that we prioritize what we get out of each of our experiences, similar to an optical-system designer creating a particular camera for different purposes, such as for a bird's-eye view or for a close-up portrait. The unique configurations of our minds allow us to see things a little bit differently from everyone else. So, our self-consistent systems may never be absolutely perfect, but they provide the foundation for our very own colorful life.

The Beauty in Imperfection

I know many people reading this book feel an aversion to imperfection. We all strive to engage in activities that will strengthen us, activities that will bring us closer to perfection. I also know that those same people feel a sense of dread whenever they don't achieve perfection. That kind of thinking leaves us rigid. We're always looking for another avenue to perfect, and we become tremendously frustrated with ourselves and others when we can't achieve that perfection. It's extremely draining to consistently work at achieving perfect results every time.

So, I present a simple solution: Stop trying to be perfect while trying to be better.

There will always be thoughts, beliefs, and actions that contradict each other in our self-consistent systems. The joy in life comes from looking for solutions to solve contradictions but realizing that they will never completely meet our goals.

So, learn to find happiness in expanding your boundaries. The failures you have are at the end of the road; they're simply a chance to try something new. Think about it this way: Failure allows you to try a new approach with new information. You have new tactics. It's like the first time you experience a classmate getting pushed down on the playground. If you speak up, the bully targets you next. But if you stay silent, there's a part of you that yearns for justice. The black eye that you might get from defending the defenseless may eventually look like a badge of honor for standing up for someone else.

This striving for imperfection also allows you to challenge your perspectives and values. So what if what you believed when you were younger wasn't true? Does that truly matter? Or is it just an opportunity to give you a better perspective? Ultimately, you're balancing your core principles. You're finding solutions to problems that work. You're evolving and growing.

Striving for adaptability hurts at any age. Stretch marks always do. But they're giving you a growth mindset and letting you view your inconsistencies as part of the learning process. By the time you're older, those wrinkles you've gained from stretching in new directions are more of a badge of honor than an ugly reflection of who you are.

Many of us are afraid to end our lives with lined faces. But what if we thought of those lines as smile marks instead? They're the signs that we developed an ever-evolving view of our personal development. They're the proof that we fought through the good times and the bad times and left behind a legacy of growth.

The Adaptability of Julia Child

For many, the prospect of getting older means winding down their careers, but that wasn't the case for Julia Child. She may now be remembered as one of the most renowned chefs and TV personalities, but she didn't start her life that way.

Julia Child graduated from Smith College in 1934 with a degree in history. She worked as a copywriter for a furniture company in New York City before joining the service during World War II. She worked in the Office of Strategic Services as a research assistant and later as a top-secret researcher in various projects, including the development of shark repellent to protect underwater explosives.[31] She had the chance to travel the world, visiting locations such as Sri Lanka and China.

After the war, she moved to France with her husband and developed a passion for French cuisine. She received formal culinary training at the Le Cordon Bleu cooking school, which became the foundation for her later work as a chef and author.

Child didn't go on a transformative journey until her late forties and early fifties. She co-authored the groundbreaking cookbook *Mastering the Art of French Cooking* at forty-nine. She began her TV career in her early fifties, becoming the beloved culinary icon known for her French style and enthusiastic teaching.[32]

Like Julia Child, we can all recognize there are different phases in our lives. For some, that's a frightening concept, but it doesn't have to be. Each of us has a multifaceted nature. The interests in our lives will always evolve. There will be points at which we don't know the direction we should go in, and there will be points of clear vision. But we'll never reach those points if we're too focused on living a perfect life. Most of us have a sense of relief knowing we can let go.

After all, that's the only way to achieve true happiness.

---

31   "Julia Child's Spy Days Included Work on a Shark Repellent," *HISTORY*, May 30, 2018. The idea was to create a method for making downed space equipment unappealing to sharks when it landed in the ocean.

32   "Julia Child's 'The French Chef' Debuts," *HISTORY*, February 8, 2024. Child was known for her zany personality that was instantly appealing for TV. She made French cooking look less intimidating for a US audience.

Chapter 7 Questions

- *What aspects of perfection do you consistently hold on to that make it difficult to adapt to different situations?*

- *How are you analyzing your priorities?*

- *In what areas of your life do you experience resistance to your personal evolution, and what stops you from looking for different solutions to your problems?*

CHAPTER 8

Putting Others in the Picture of My Experience

Thus far, this book has been about personal journeys. We've discussed individual lives and how you can achieve your goals, so this chapter may seem out of place. It's all about living with other people and putting them in the picture of your own experience. Though this chapter is just one of eleven, it is, in my opinion, one of the most important.

I've seen a disturbing trend in online communication and communication in the workplace. So many of us currently work remotely, and most of our communication is through text. As a result, there is an empathy crisis. Many of us have learned to internalize all our opinions, whether we deem them acceptable or unacceptable. We're afraid to raise our voices and agree or disagree with friends, though the expression of our differences in opinion is so vital. It's a rather sad state of affairs. Our differences in opinions can drive creativity and diversity of perspectives that lead to innovation. They're vitally important, but they're disappearing.

Our individual liberties are limited by other people, as every individual, you or anyone else, is entitled to seek freedom. We don't live by ourselves. We live with others. We cannot exercise our freedom at the cost of others'.

I like to compare it to drivers sharing the road. If you look at the vast network of roads in each country, you'll see just how far all of us can go. But we do have a limitation on our motion on the road itself—to the left and right are lines that instruct drivers to stay in individual lanes. If you're moving along a four-lane road, and you're traveling parallel to another vehicle moving in the same direction, you can't move over to occupy their space. You'll either crash or force them off the road. For both your safety and theirs, you must respect where you each are in space when you're trying to get where you want to go.

We live in a community with others. We each have the liberty to speak our minds, and we each have different opinions based on our own experiences, but we'll never truly get anywhere in life if we don't have an eye set on our communities.

I'll put these questions to you: How are you putting others in the picture of your experience? Are you using empathy to see someone else's perspective?

We each need to form a nonviolent communication style that assumes people are, by nature, like ourselves. If you are compassionate, they can be compassionate as well; if you are not, neither might they be. Your promise not to drive your neighbor off the road when traveling on the highway relates to them as well. You have a lot in common. You're both moving forward in life with a social harmony that accommodates each other. It's a sense of fraternity that allows you and the people around you to flourish.

The Age of Reflection

A community of spirit and mindset thrives throughout our experience. We cease to exist without one another. Our societies thrive on the belief that the prioritization of people is one of the most valuable assets we possess. At various points in life, you experience the childlike connection between two people who have nothing more in common than a favorite color, then deep fraternity bonds, and then family ties.

And yet, there is a point at which we stop and reflect on what those relationships were like. We see their importance and how they've shaped us. They show us new things about the nature of humanity and ourselves.

As we continue this journey along life's track, we reach an interesting point in humanity's timeline. There's a lot to be said about turning the corner into retirement age. Our priorities have changed, and we no longer look for a simple paycheck to satisfy what we need. It's a time when most of us start to prioritize family and friends more than we ever have. At retirement, we may have more time to make sure that people come first. So, let's take a deeper dive.

You find yourself sitting with the community council as they decide on a new parking lot for your town. It's not something that you would have ever done before, but you're trying to make a difference. You've recently learned about the rarity of a certain species of owl, and the new parking lot would destroy some of its habitat. It's the first time you've ever advocated for an owl, but you've seen it before and grew up with it, and it would be a shame if it disappeared before your great-grandchildren had a chance to enjoy it.

You lean over to whisper to your spouse, who came along with you to support you. Your spouse has a few friends who are wildlife advocates, and they want to put in their two cents if you're going to be the first one to speak.

When the council asks if anyone else has anything to add, you stand up, notebook in hand, and make your way to the front of the room. You tell them of your newfound interest, and you're hoping that they see the value in the wildlife that might lose its home. You mention that you know many people who have the same opinion, which is why you've made your way there today. You want to be the spokesperson for everyone in the community who has a vested interest. You motion to your spouse to bring up a deck of slides that you've prepared, showing the potential damage that could occur with the new parking lot.

You breathe a sigh of relief as you make your way down to your seat again. The council seems to be in your favor, and you know you've made a difference.

You glance behind you to see the people who've gathered to advocate for the parking lot. "One moment," you say as you turn again toward the council. "I might actually have an alternative for that parking lot. A friend of mine owns a construction company that frequently works with large buildings, and I know for a fact that he has helped put in a few parking garages. Our parking lot down the street could use some revamping anyway."

As you make your way to your car, you hear someone yell, "Hey!" One of your spouse's friends is beckoning over to you. You cautiously look around you before heading over to talk to her. She appreciates what you did in there. Then, she asks if you are free to head to the community center. She and a group of other people have organized a bit of a party in response to your bravery in standing up for the owl species. She wants you and your spouse to join.

You stand and stare for a moment. It's the first time you've ever been invited to something like this. But you've decided to make it a priority to reach out to the community more, and this is an excellent start. You nod your head and smile as you walk with her, your spouse, and a community of like-minded people to the community center.

The tendency to put ourselves above others isn't contingent on age. We often are less accommodating to new things in other people's behavior, regardless of how old we are. And yet, it's vitally important that we put other people in our picture.

We've all experienced interactions with unpleasant people. If our own feelings are hurt, it's easy to put other people's feelings aside. But what if we instead understood other people's frustrations? We're asking other people to have empathy toward us, so it's necessary to empathize with them. We all rely on one another, which is why we need to retain that sense of empathy with everyone.

Of course, learning to do so is difficult. It usually takes a lifetime to learn to calm our minds and look outside of our own perspectives. The way we envision the correlation of our perceptions with reality may differ wildly from someone else's, and yet, their perceptions may hold a key to helping you better correlate with reality.

From Reliance to Inspiration

Helen Keller's life began in darkness. At just nineteen months old, she lost her sight and hearing because of an illness. She struggled to make sense of her surroundings, and her intense frustration with her inability to communicate led to frequent outbursts. Her family was unsure of how to help her, and the solutions at the time were limited.

When Keller was six, Anne Sullivan was hired as her teacher. Sullivan came from the school for the blind. She showed extraordinarily patient teaching methods to help Keller learn how to read. She began by spelling words on Keller's hand, a method that was initially met with stubborn avoidance. Her first breakthrough came when Keller connected the sensation of water on her hand with the word *water*.[33] It was a light-bulb moment. Suddenly, Keller's realization that she could understand language and communication ignited her passion for learning.

Keller went on to master several languages and graduated from Radcliffe College, living as a passionate advocate for people with disabilities.[34] She had so many obstacles in her life, and yet, she overcame them all. It was revolutionary.

Because of her own experiences going through the difficulties of disabilities, she understood the need to help others transcend their own limitations. She dedicated herself to the service of others, connecting with people from all walks of life and fostering a sense of community and shared humanity. It was a rich life that promised meaning and impact. It was the chance to empower people and contribute to the greater good.

Helen Keller's story was groundbreaking for its time. Most people had prevailing negative attitudes toward people with disabilities. For many, living life with these disabilities meant being barred from success. But, with Keller's advocacy, attitudes changed. People with disabilities—those who had accessibility, education, and empowerment—could create their own personal freedoms.

Helen Keller's legacy has far outlived her. She laid the groundwork for the Americans with Disabilities Act, which passed in 1990.

---

33 Helen Keller, *The Story of My Life* (American Foundation for the Blind, 1903).

34 "Helen Keller Graduates from Radcliffe, First DeafBlind Person to Earn a B.A.," *HISTORY*, March 14, 2024, *https://www.history.com/this-day-in-history/june-28/helen-keller-graduates-radcliffe-first-deafblind-college-grad.* Keller's swift adoption of communication gave her the hope to inspire others.

It prohibits discrimination based on disabilities and provides equal access to public spaces, employment, and services. The Helen Keller International organization,[35] cofounded by Keller herself, was formed to combat the consequences of blindness and malnutrition worldwide. Special education programs and inclusive teaching methods—such as screen readers, Braille displays, and communication devices—were inspired by her work. She also inspired people from all backgrounds to take up the mantle of advocating for disability rights and inclusion. It's a legacy that has lived on for over a century and will continue to live on for many years to come.

A Vision of Others

It's remarkable, the difference that one person can make. By providing the avenue for Helen Keller to learn, Anne Sullivan led the charge for Keller to become an inspiration to millions. Her simple act of compassion was the foundation for an entire movement.

It should go without saying that including people in the scope of your life is a vital way to grow, but there's more to it than that. We depend on each other for everything. The lives we live today have been made possible by the sacrifices of those who love us. It's only through our empathy with others that we can truly become the best people we can be.

We grow up believing in the individual. We see the world from our own perspectives, seeing ourselves at the center of the world because those are the only eyes from which we can see. It's almost as if you are living independently from reality. Though you may know

---

35 "Our History," Helen Keller Intl, accessed November 25, 2024, *https://helenkellerintl. org/who-we-are/our-history/.* Keller traveled the world to advocate for people with disabilities for the rest of her life.

that you are but a small person spinning on a rock orbiting a sun in one galaxy among countless galaxies, you are at the center of your own story. To you, the lives of others are only secondary parts of your overall story.

We are all who we are because of the decisions we have made. Everyone has the freedom to act as they will, to explore who they are. The freedom of individuality urges you to go outward with no limit, to explore the boundaries of what you can do.

But freedom has limits.

Understanding Equality, the Boundary of Freedom

Equality is fundamentally about ensuring that every individual is treated the same as the next individual, no more and no less. While different individuals may seek different rights and different opportunities, all individuals have the same rights and opportunities as each other if they seek them, unless they are freely willing to give them up or trade them. That also means that one person's freedom shouldn't come at the expense of another's freedom. None of us is superior to anyone else, and our society should reflect that.

Imagine, if you will, a farmer living on the same property his family has owned for a hundred years. The other farmers eventually sell off their land for development, and a local HOA is enacted. The HOA is frequently annoyed by the farmer's need to fertilize and care for his crops, and the farmer is frequently fed up with the HOA's desires to take his land and enforce restrictions on his livelihood. While the parties are separated by a fence, their two extremely different lifestyles clash. However, their personal freedoms cannot interfere with each other.

Over time, both sides learn to live with new lifestyles. They form boundaries, but they ultimately learn to respect the other's existence as equal to their own.

Our need to coexist is much like a waterfall. As individual water droplets cascade over the side of a cliff, many of them hit rocks or outcroppings on the way down. The water flows fluidly around each obstacle. The water droplets have the freedom to drop and flow, but there is also a boundary to somewhat limit their free flow. Finally, the water droplets that land in the pool below collect and create a lake. Over time, a new balance is created for coexistence.

Each new water droplet that tumbles over the waterfall exercises a new freedom and a new boundary.

To live a full life, we need more than the contributions of others; sometimes, we even need sacrifices (when others are freely willing) to prop us up. Likewise, we may contribute our efforts, and sometimes sacrifice ourselves, to prop up others when we think it's worth doing so based on our own value systems. We all desire to live in a reality where we can coexist peacefully with the people around us. But achieving that requires a direct consciousness of our self-awareness.

There's an ethical problem that develops when we don't have self-awareness. It's necessary to look outside our own scope to someone else's. The moral high ground should involve equality, the treatment of others that equals our own. So, one individual should not exercise a freedom that will hinder someone else's freedom. This is realized by humankind through the evolution of human civilization. A society that prioritizes someone's personal freedoms above those of others has to be dependent on power and privilege. And the rise of individualism is a critical part of a modern civilization in which the underprivileged are respected and the voices from the weak are heard.

The proper response is to cultivate a culture of empathy and mutual respect. We're all interconnected because of our freedoms. Fundamentally, understanding your purpose in life helps you express liberty without doing so at the expense of others. We need to create a balance that allows everyone to thrive.

Everyone has the right to push the boundaries of their personal freedoms, but not at the cost of anyone else's. In fact, to make sense of life, we need to explore our unlimited potential and pursue our aspirations.

The boundary of including others in our own understanding of life is flexible. It's a line that evolves and progresses as we expand our horizons. Of course, that means that understanding where the balanced boundaries are requires constant analysis. With empathy, we can base that analysis on the pursuit of personal freedoms, principles of fairness, and mutual respect.

Developing a Fraternity

When many of us think of the word *fraternity*, it's often associated with universities and high elites, but that's not where it originated. Fraternity, in its essence, refers to a sense of community, a solidarity among individuals. It's the bond we share that connects people to a sense of unity and mutual support. While freedom is more focused on individual desires, equality comes when two or more individuals form a fraternity focusing on a community. Fraternity is based on trust, respect, shared values, and creating a network of support. At its heart, fraternity is empathy.

If we're to create a true sense of meaning in life, we need this sense of empathy. It's the key to preventing isolation.

We often describe mutual connection and shared understanding in terms of putting yourself in someone else's shoes. When was the last time you listened actively to someone? When was the last time you shared your perspective and looked for another perspective in return? We all want to feel heard and understood. But that feeling goes both ways. Have you ever made efforts to listen to others?

Often, we're wrapped up in our own motivations and feelings. There's nothing wrong with recognizing what we want, but doing so at the expense of others is a drain on us emotionally and psychologically. The more we take others for granted, the less we grow ourselves.

It's quite strange what empathy does to you. When you're young, sharing your toys with someone else seems like a burden. It's something your parents tell you to do, but you hold a slight resentment, no matter what. But there's wisdom in that empathy, in that shared community. When we start to think outside of ourselves, we become infinitely more adaptable. You're no longer simply thinking of avenues *you* can take; you start thinking of directions *we* can take.

It's far less lonely taking this path with other people, even if there are personal sacrifices. But that's the point, isn't it?

Humans are very social creatures. We rely on each other to bolster each other when we feel down and to create a sense of community. Evolutionarily, we relied on each other to survive. We formed social groups to better hunt, gather food, and protect ourselves. These are mechanisms that we've continued to embrace over the years. In fact, the prefrontal cortex and the mirror neuron system were each designed to help us better interact with each other. We receive a healthy dose of oxytocin during social interactions, which further reinforces bonds and promotes trust and connection. In essence, it's such an important part of us that our minds literally evolved to support each other.

So, the support we demonstrate for each other is necessary for survival, even in a world built around modern conveniences. At some point in nearly all of our lives, we say that we can do it on our own.

But we always return to the comforting sense of community we've built. It's natural, and it's beautiful.

But we can go a step further.

One of the most valuable skills we can possess is associating our happiness with someone else's. The natural instincts in us make us want to believe that we can only feel happiness when our needs are met. But your success multiplies significantly every time you value someone else's happiness together with your own. You're growing your team's tank. You're filling it with fuel that will take you further than you could have ever gone yourself.

A Community of Diversity

One of the greatest evils sold to us nowadays is the belief in sameness. I'm not just talking about people who look the same or grew up in the same town. I'm talking about the value of prioritizing different perspectives. It's easy to become polarized, to create a "them versus us" mentality. In fact, that kind of thinking has existed for as long as we have. It's also the basis for fun activities, such as sports. But at a fundamental level of cohesion, of living together harmoniously, it cuts at the very purpose of creating a community. Even in sports, sportsmanship is promoted in modern civilization. You may believe that you have created a solid fraternity among the people who hate the same rivals you do, but this is merely the first step in division. In those cases, we're again focused on the individual only, not the community.

There are many examples of this, but let's go back to one that is known by most people: the French Revolution.

In 1789, the French population highlighted a severe schism when the French elite put heavy taxes on the majority of the population. Many of the elites held political power and took the money of those less fortunate. There were significant hardships in the form of poor harvests and rising bread prices leading up to the revolution, which furthered the distance between the upper class and the lower class. The government was in deep debt—partly because of wars overseas, which included the American Revolution—and the lower class was taking the brunt of it.

At the time, there was a voting system that granted everyone a vote. Yet, the lowest social class had such a small say in what actually happened, despite their larger size, that they were frequently outvoted. Feeling that they had no other choice, the lowest class, called the Third Estate, formed a National Assembly to advocate for equality, liberty, and popular sovereignty. When they brought these grievances before the ruling class, they were met with laughs.

The lower class became so frustrated with the little power they had that they started thinking about a violent revolution. If the monarchy and the nobility didn't care about the common people, they would soon feel their pain.

When the peasants rose up to overthrow the monarchy, they were further enraged by the riches they saw when entering the palaces. They dragged the monarchs from their homes and imprisoned them as an act of defiance.

In the resulting vacuum of power, radical political groups rose up and began executing anyone they perceived as a threat to their revolution. The constant shifts in political power meant that there was no stable economy. No one felt as though they belonged to a community that was safe enough to live a normal life. Eventually, it fell upon people such as Napoleon Bonaparte to seize power and once again create a ruling class through war.

The deaths and chaos that came as a result of the French Revolution are some of the most brutal in history. And it all started because of too much focus on the individual.

Recognizing Freedoms

It's not enough to simply acknowledge the differences we see among ourselves. If we start seeing ourselves as the only people who matter in the world, we start to lose humanity. It's at that point that we start losing our own freedoms as well.

We live lives that are dependent on other people, whether we like it or not. If we restrict the freedoms of others, we're restricting our own freedoms as well. If, for example, we cut ties with people who were once in our group of friends, we lose the strengths and abilities they brought to the table. Suddenly, we're not given as much freedom as we thought we were.

To combat this problem, we need to understand that not only should we recognize others' freedoms, but we also need to try to understand the world from their perspectives. Instead of simply acknowledging that our points of view are different, we should actively seek out the other point of view. More often than not, there are other perspectives that are much more valuable than we might imagine.

Building Trust

Building trust is one of the most difficult parts of creating a community. For those of us whose trust has been violated, it's even more difficult. For many, it's often seen as a risk not worth taking.

Let's take a rather extreme example. Serial killers are more likely to attack people who are willing to give them a hand. If you stop on the side of the road to help someone who's changing their tire, you might be met with a knife. But the chances of this happening are extremely small. In most of those cases, giving blind trust to help someone out will pay off for you, one way or another.

In the vast majority of cases, trust becomes the cornerstone of living a happy life. But here's the most important part to consider:

You need to earn trust from others with your own efforts but also try to give others your trust when you can afford the risk of that trust being violated.

If you're giving trust to someone, you're on the same side. You'll start to see that you're choosing to see things from someone else's perspective. You're choosing to acknowledge their realities and accept that their perceptions may be flawed. You're not giving your trust blindly. Just like with every other aspect of your life, you need criteria.

There's a risk that your trust may be violated by anyone. Anyone could steal from you, lie about you, or share intimate details about you with someone else. When you first start a relationship, the chances of something like that happening are small. These small risks are well worth it when you're first developing trust. The small tasks that you work on together build cohesion between you that could blossom into something much more beneficial for both of you. Over time, what started as surface-level trust could blossom into something that could withstand fire.

Once your relationship gets deeper, you need to start using your judgment. Maybe you can't afford the risk of telling the other person a secret. Maybe you can't even afford to let them drive you home. As you evaluate each relationship, you can come up with criteria on the fly that specify how well you can trust people. I like to think of it this way: If you can afford the risk, give your trust. If you can't afford the risk, then that trust is blind.

If you're all about the numbers, it's pretty simple. Most of the time, it's easier to give your trust, if only on a surface level.

But there's much more to it than the benefits other people get from receiving your trust. Ultimately, you'll be happier if you trust other people. Constantly questioning how other people feel and what their motives are is eternally exhausting. Everyone else has perceptions and motivations different from yours, so if you live your life believing that no one else has your best interests at heart, there may be some truth to that, but it shouldn't stop you from trusting them.

The negativity you feel about other people will only spread. If you constantly believe that other people are out to get you, your anxiety will grow. You'll also start to feel more paranoid over time as you struggle to reconcile those feelings.

What's more, you'll end up gaining much more from your life if you focus on building a community with other people. You never know what you're missing by being too aloof. Yes, you might save yourself some heartache, but we thrive so much more when we rely on others. We're suddenly looking outside of ourselves and growing along the way.

A Bond Forged in Iron

Larry Page, one of the founders of Google, grew up in rather extraordinary circumstances. He was part of a computer-heavy household and was taught computer programming at a young age, as his father, Carl, had earned a PhD in computer science in 1965.[36] It became only natural for him to pursue the same career as his father.

In the same year, Sergey Brin started his life in a considerably different place. His family originated from Moscow, Russia, with both parents coming from academic backgrounds. The family left Russia in 1979 to make their way to the United States.[37] Brin started his career

---

36   "Larry Page," *Academy of Achievement*, accessed December 17, 2025, *https://achievement.org/achiever/larry-page/*.

37   "Sergey Brin," *Biography*, last updated March 26, 2021, *https://www.biography.com/business-leaders/sergey-brin*.

in academic excellence in 1990, starting his BS in computer science in 1993, which is when he met his counterpart, Larry Page.[38]

Their connection sparked something they could have never created without each other. Their time in computer science convinced them that they needed to find a way to make accessing information much easier. Page brought web experience with his electrical engineering degree, and Brin brought expertise in data mining, computer science, and math.[39] The combination of the two made it possible to create a data search algorithm that they named Google, after the mathematical googol, the one followed by one hundred zeros.

Their success blossomed into one of the most successful companies today. But, perhaps most importantly, it illustrates the point we often forget in our own experiences: The only way to succeed is to challenge your worldview by experiencing new things from new people.

Page grew up in a family with abundance, never having to abandon his dreams of pursuing computer science. Brin, on the other hand, set the foundation for what would become Google's famous motto: "Don't be evil." The idea was to combine free information with a commitment to healthy values to create an open information source anyone could access.

It's extraordinary how simply adding people to your community can make such a large difference. The strengths you have on your own can only get you so far. Many of us have experienced what the value of a good friend, a spouse, or a child can do to improve our lives and take us much further than we could have gone on our own. Who we are, in fact, is often a blend of everyone around us.

---

38   Uri Daigin, "Sergey Brin—Biography," *JewAge*, accessed December 17, 2025, *https:// www.jewage.org/wiki/en/Article:Sergey_Brin_-_biography.*

39   Sergey Brin and Lawrence Page, "The Anatomy of a Large-Scale Hypertextual Web Search Engine," *Computer Networks* 30, (1998): 107–17, *https://snap.stanford.edu/ class/cs224w-readings/Brin98Anatomy.pdf.*

We develop a sense of understanding of where we are in our lives because of the experiences we've had, but that's not the end of the story. At various points in our lives, we've become stronger together with people in our communities. Our freedom isn't hindered by others but rather grows because of them.

Chapter 8 Questions

- *What significant changes have you seen in your life because you were willing to add someone else to your community?*

- *What personal freedoms have you sacrificed to respect another's freedoms, and can you find anything you may have gained from doing so?*

- *How has your worldview changed by experiencing things from someone else's perspective?*

PART III:

The Match for a Happier Life—Your Experience and Your Desired Experience

CHAPTER 9
Know Yourself

By now, you've likely developed a clearer picture of how to make sense of your own life. The philosophical climates taken to get here require serious personal effort. You're developing a sense of who you are and the personal journey you need to go through to become whom you want to be. (I, of course, am speaking in terms of continued development. Even philosophers who had all day, every day, to think about their lives might never have completely reached the pinnacle of their potential.)

But, moving forward, I want to focus more closely on leading a happier life. The correlation between our vision of reality and reality itself needs to be a good match to ensure we are living as happily as possible. At its heart, that means aligning our desires and our skills with what will truly make us happy.

So, how can we make what we desire and what we're good at match? As Marsha Sinetar said, "Do what you love and the money will follow."[40]

---

40   Marsha Sinetar, *Do What You Love, the Money Will Follow* (Dell Publishing, 1987).

Now is the time to put those theories and axioms to the test. Where have you found that your skills and knowledge help you build confidence and serve as a foundation for your true desires? Typically, those desires are matched with your expertise and the benefits that you give to others. For example, if you spend your time frolicking in the woods with an eye out for all of the indigenous nuts, fruits, and herbs that line the forest floor, you may be an herbologist at heart. If you find yourself constantly daydreaming about running marathons or karate-chopping bad guys in the street, you may be a comic book writer or a ninja turtle at heart.

It's a beautiful thing to find what you truly desire. And it all starts with that crucial word: *I*.

It's common to feel like you're lumped in with a crowd. If you have an aptitude for math, maybe you'll join the army of accountants. If you have a desire to dance, maybe you'll join the ranks of ballet dancers. Whatever the case, I can guarantee that, at least at one point in your life, you were told you had to do something based on an aptitude or desire you possessed. You were told to take a test to find out what you should do in life because you were "good at something."

I hope I'm not the first person to tell you not to take those guidelines as seriously as you might want to. I know that many people I've talked to have taken years off their lives conforming to what they believed someone else wanted from them. They've pushed away true happiness because what they wanted wasn't in fashion or wouldn't make them money.

Only you truly know yourself. You may be good at math, so someone told you to be a math teacher, but what if your true desire is to draw? You may be told you have an aptitude for speaking, but what if you want to own a farm instead? None of the options should be limited to you because someone said you should choose one direction over another. Your goal should be to match your desires with reality instead of forcing yourself into a box.

Though this type of advice is typically given to people starting out, it's just as poignant at any point in your life. Perhaps you didn't start going after your desires in your early career. Sometimes, our desires change and waver over time, but there's never a better time to seek what you desire than now.

The Seeds of Retirement

On our journey through life, we often associate the early senior years with an abundance of time but a loss of most things we've come to know. It's when we actually have the time to do some of the things that we've always wanted to do. It's why you'll see new retirees with sports cars or engaging in some activities they never had the time to do before, such as cliff diving. So, it's appropriate that we reach this phase at this part of this book.

Imagine, if you will, the time after retirement.

You're setting off for the first time in over a decade. You've bought a new sailboat, one that your spouse only begrudgingly agreed to purchase. You've always wanted one, and now is the time to make that happen. It's been a part of your dream since reading *Robinson Crusoe* and wanting to sail around the world. Of course, you know the sailing-around-the-world part won't actually happen. But that doesn't mean you can't dream about it from the shiny railing of your new boat.

As you step onto the bow, you feel a slight crick in your knee. It's a familiar sensation. You've been needing to get your knee replaced for some time now, but you've never really wanted to go through with it. At that point, you'd have to accept that you are getting old. And that's just not something you want to acknowledge.

You hear a noise behind you, and you turn to see your grandson, yelling your name. You smile as you see him running up the dock to you. You jump off the boat, throw your arms around him, and lift him into the air with a tight squeeze. "How're you doing, Tiger?" you ask when you set him down and ruffle his hair.

"I get to go on the new boat!" he exclaims.

"Not only that," you say, "but you might even get to steer the boat, if you're up to it."

He lets out a squeal and runs for the boat's helm. It makes you chuckle. You remember when you were that age, and you would have loved being in the position that your grandson is now. You take a minute to stare at him for a little longer before turning to your daughter, who is coming in for a hug.

"I've missed you," she says.

"I'm so glad you made it," you reply.

There comes a point in everyone's life when their priorities change. Some people become more important than ever, and there are relationships you can't just let go of. You start to realize that the way you're using your time and efforts makes a big difference. You no longer want to devote your time to things that don't make you happy.

If you're lucky, this realization will come sooner in your life rather than later. It's something that often comes with blossoming wisdom, which is why a lot of people don't realize what they really want until later in life.

But being happy isn't just about acknowledging your strengths. More often than not, it's more about learning to let go of your weaknesses. It's about accepting who you are no matter what. And that is truly one of the most difficult realizations you will ever make.

Flying Through Thought and Action

Born in 1452 in Vinci, Italy, Leonardo da Vinci didn't have the best start. He was the illegitimate son of a notary and a peasant woman, and he received an informal education in Latin, geometry, and mathematics because of his lack of resources.

He also showed an affinity for art. In a twist of fate, he became an apprentice to the renowned artist Andrea del Verrocchio in Florence.[41] It gave him the opportunity to gain skills in painting, sculpture, and metalwork, something he might not have had access to if his father hadn't recognized his talent early on. It was an immense opportunity that gave him the desire to expand to a variety of skills in art.

He created detailed sketchbooks with a deep understanding of light, shadow, and human anatomy. They inspired masterpieces such as the *Mona Lisa* and *The Last Supper.* They were the perfect blend of science, engineering, and creative passion. Though to the untrained eye, they would seem only paintings, each of them displayed an acute knowledge of geology, optics, and hydrodynamics.

Maybe it was this unique fascination with science that inspired him to reach outside of mere paintings.

When Leonardo looked to the skies, he saw a unique blend of artistry and mechanics in the form of bird flight. He sketched detailed images of wings and became fascinated with the idea that humans might also one day fly. It became an overwhelming desire that later emerged prominently in his writing. So, he attempted to create several

---

41   *Britannica,* "Leonardo da Vinci," by Ludwig Heinrich Heydenreich, last updated December 9, 2025, *https://www.britannica.com/biography/Leonardo-da-Vinci.* Verrocchio gave Leonardo a multifaceted training in everything from art to technical-mechanical designs.

flying machines based on his sketches.[42] Sadly, none of them worked. Leonardo da Vinci was acutely aware of his lack of resources. He could make his machines glide, but none of them truly flew. It became much more difficult to construct and test flying machines as he grew older, so his work diminished with his age. Still, he never truly gave up on the idea of a flying machine.

It wasn't until four centuries had passed, in 1903, that two engineers would make Leonardo's dream come true.

Orville and Wilbur Wright extensively studied Leonardo da Vinci's sketches.[43] He had designed an ornithopter, a machine intended to mimic the flapping wings of a bird, but he'd never been able to make it work. The Wright brothers saw the complicated engineering design and adapted it in their analysis of lift and thrust, critical components of flight. The wings were designed from his sketches, and the experiments with wind tunnels came from direct inspiration from Leonardo himself. They also built control systems for the aircraft that mimicked Leonardo's designs. Centuries after Leonardo da Vinci drew the original designs, they had the resources to make a real attempt at creating them.

A successful flight took place at Kitty Hawk, all thanks to the inspiration Leonardo da Vinci's notebooks gave the Wright brothers.

In his heart, Leonardo probably knew that he would never live to see human flight. It was a result of his situation in life. Though he never truly gave up on the idea, he long knew that he was bested.

---

42   "Leonardo Da Vinci's Dream of Flying," *LeonardoDaVinci.net, https://www.leonardo-davinci.net/flyingmachine.jsp.* Leonardo studied many birds and created his *Codex on the Flight of Birds* in 1505.

43   Scott Johnson, "History of Engineering: Case History of Flight," *LibreTexts,* November 9, 2020, *https://eng.libretexts.org/Bookshelves/Introductory_Engineering/ EGR_1010%3A_Introduction_to_Engineering_for_Engineers_and_Scientists/11%3A_ Historical_case_studies_in_Engineering/11.01%3A_First_Flight.* Interestingly enough, both were fascinated by the idea of helicopters, but neither of them would see that vision become reality.

What Drives You

Many reading that history may have felt Leonardo da Vinci's loss. He would never fulfill his dream of flying, even though he gave years of his life to the attempt to do so. But there's actually something beautiful in the result. Leonardo gave his life to the pursuit of developing new ideas. He challenged what he believed was possible and came up short. Ultimately, he couldn't make his ambition match his reality.

But his dedication to his work gave others the opportunity to continue what he had started. The Wright brothers lived in a time when resources allowed them to complete their vision. In the end, Leonardo da Vinci didn't waste his entire life believing he could achieve what he never could. He laid the groundwork for others to achieve what he wanted.

Another interesting outcome of this mismatch was Leonardo's devotion to a multitude of other subjects. When he realized his limited resources would never result in a flying machine, he spent the last years of his life refining the ideas he'd accrued. He learned far more about anatomy, hydraulics, and mechanics, filling notebooks with sketches and observations and exploring topics such as the movement of water and the human body. Though he would never see the publication of his studies in his lifetime, his insistent organization made it easy for others to publish them after his death.

You, reading this book, might be startled by my suggestion to focus on what you're good at. Most people are told from a young age that they can achieve anything they put their minds to. It helps to inspire young people to chase their dreams at the beginning while they are exploring, but that's not necessarily the case when they actually grow older. (The underlying reason for this change is that younger people have more time to risk finding something they might not like, while older people have less time to waste. This itself is an example of criteria change following scope change.)

It's time to match what you want with what you're really good at.

Knowing What Drives You

A lot of people let their passions slip by because they can't articulate them. I find that most people, at some point, get very frustrated because their passions aren't speaking to them as loudly as others' visions are to them. That dreaded question, "What do you want to do with your life?" very rarely truly answers the question of where your passions lie. Most people think of a job, something they *have* to do to get by. But your passions are much more likely connected to the top five things you're always doing.

Let's do an experiment. At the end of this paragraph, close your eyes, clear your head, and sit back in a chair. Don't fight the first thing that comes to your mind. Embrace the thoughts that you keep coming back to. Take at least five minutes.

Now that you're back to reading this book, answer this question: What did you think about?

We're often most passionate about the things we think about the most often. If you're constantly thinking about a podcast you want to start, an engineering project you're dying to try, or a sport you just can't get enough of, you've identified your passion.

If you're still struggling to find what you're most passionate about, you may find that you have a lack of interest in yourself. Finding out what you truly want takes time, and that means spending hours in silence listening to yourself. Your passion is an extension of who you are, and the only way to truly understand yourself is to take a deep dive into your passions. What you spend the most time doing speaks volumes about your passion. Where do you dedicate your time not spent in obligation? You might be surprised to discover what that passion really is.

Knowing yourself also means understanding your purpose. What are you good at? If something comes easily to you, it could be a good sign that it's matched with your purpose.

Knowing the Mismatches

Of course, not all of the passions you might desire fit with who you are.

Elements outside of your control determine these mismatches. If it's in your control to grow and develop your passion, hold on to it. If it's not in your control, give up on it.

Some people spend their entire lives chasing after something that is obviously beyond their reach because of physical limitations, and many enjoy the process, even though they know they won't get to the final point. Some people with debilitating asthma desire to climb Mount Everest. Others who have difficulty paying attention may want to be chess grand champions. These are worthy goals, but they're ultimately beyond their physical capabilities. There's a mismatch between what they want and what they can realistically achieve.

In the end, everyone is forced to take an eternal break. What's not in your control may just waste your time. There are so many other passions you can develop that are within your control. It would be a shame if you were the world's best writer but spent your entire life as a mediocre cook.

The downside of discovering these mismatches is that they often hurt. When you're tired, sometimes you just need a break. Sometimes you just need to say, "I did what I did, and I did my best, but I can't do it anymore." It's a sign that you're pursuing the wrong things, the things that are pulling you down instead of pushing you up.

Some things are simply outside of your control. What if a passion you have is more detrimental to your health than it is a saving grace? The solution to this heart-wrenching push? Don't forget, you can always put it in God's hands at any point when you feel powerless.

It is interesting that many people embrace religion as they get older. Isaac Newton, though religious throughout his entire life, began intense theological study in the latter years of his life. C. S. Lewis, initially an atheist, converted to Christianity later in his life and became one of the most influential Christian apologists in the twentieth century. Carl Jung, a psychologist, delved more deeply into the spiritual and human psyche, often reflecting on a complex relationship with religious ideas. All of these people were deeply invested in what humanity could achieve, but as they got older, they found peace in putting it in God's hands.

There is only so far you can make it on your own. Whom you believe in toward the end of your life, as you start to feel your own strengths failing you, may allow you to simply let go. It's an act of faith that lets you put your trust in and give control to something outside of you. You can only push the boundary so far.

Knowing Your Expertise

Once you've separated your mismatches from your passions that will sustain you mentally, emotionally, and physically, you can pair those passions with expertise. The right match between you and your passions makes you far more competitive and gives you an edge over everyone else.

Steve Jobs, a co-founder of Apple Inc., had a deep passion for technology, design, and innovation from a young age. His work with electronics and computers, paired with hundreds of thousands of hours refining his expertise, led to one of the most influential technology companies in the world. He left a legacy of blending both creativity and technical skill to create new and innovative technology.

Jobs knew from a young age what he wanted to do in life. He devoted his time to his passion and created an empire. It took years of trial and error, struggle, and sacrifice, but he rose above the rest.

It's easy to look at an example like that and think you are way behind the curve. Maybe you think you don't have any expertise. But the beauty of passion is that you can create that expertise. The purpose of your expertise is to help you follow your desires and passions. It's never too late to develop a new expertise. It's never too late to start something new.

Matching Your Passions and Expertise for a Financial Return

The world is constantly changing. Perhaps many of you grew up thinking you would have the time and the money to do the things you've always wanted to do, but that's not always the case. The harsh reality of life is that we need to make money. Freedom takes cash.

But there is something that most of us get wrong, at least when we're worried about how we're going to have time and money for everything. Though it's much easier to spend as much time as you want on whatever you want when you have money, you don't need to be rich to exercise your freedom. But you will need resources.

That looks different for everyone. Many people choose a career that keeps them close to their passions with a minimum income. For example, some people strive to have an adventurous lifestyle with constant thrills both in the air and on the water. These people might choose to live as zip-line instructors in Costa Rica, where they can also kayak whenever they want. They won't make a lot of money doing it, but the resources provided to them make this lifestyle possible. For others, their dreams could be as simple as wanting to live in a Manhattan apartment. It takes more resources to make that possible, but they can afford it by budgeting correctly, using public transportation, choosing more affordable housing programs, and networking. It takes a little more stretching, but it's possible to make that dream happen.

The key is to set up a successful monetary draw through which you can fund whatever lifestyle you want. Unless you already have all the money in the world, you'll need to tie your drive and expertise to your financial resources.

The first step in effectively matching your financial returns with your passions is to identify your target customer. Everyone has someone they report to. Whether you have a job or own your own business, you need to have a deep understanding of who would benefit most from your expertise and how your passions align with their needs.

Do research to gather insights about potential customers. Include their demographics, preferences, and spending habits. You're getting to know your customers from the inside out. Once you know exactly what they want, you can tailor your offerings to meet their specific needs. You can tweak your passions and your drive to match those of others, making sure that you have an endless supply of customers to keep up with your lifestyle.

If you are an employee, discover what value you can bring to your employment. Align your vision with your employer's to get the most satisfaction from life. If you find your values don't align, maybe it's time to find another job.

Create a value proposition in a clear statement that explains how you solve your customers' problems or improve their situations. For a business, every product or service it has should have unique benefits and value, and it highlights those benefits when it talks to its customers.

So should you. Identify how you can connect your expertise with your customer's benefit and make them understand and value it.

The way you engage with other people ultimately sets you apart from your competitors. Are you talking to them in a way that shows you're deeply connected with their problems? Oftentimes, if you share your passion with your customers, and your passion can benefit them, they will loyally follow your company. They also want to engage with like-minded people who share their passions.

Let's say, for example, you have a passion for creating digital art. If you can explain why someone needs digital art in a way that piques your customer's interest and shows that you share their values and passions, you're more likely to connect on a deeper level. You're creating a community, a fraternity.

It's this final point that I find the most valuable. You're creating a system to display your authenticity. You're not changing who you are to meet the expectations of someone else. It's much more valuable to stay true to who you are and benefit others because of it.

A Glance into Weakness

There's so much to devour when analyzing your true self. Knowing yourself is a never-ending journey, something you don't fully realize even with your last breath. And perhaps one of the most difficult aspects of our existence is the awareness of our weaknesses. We all have them, but they're extremely difficult to acknowledge. Most people shroud their weaknesses in a good layer of strengths, putting as much distance between them as possible. But they're always there. No matter how much you accomplish, no matter how much time you spend trying to strengthen the cracks, you will always have weaknesses.

If you're one of those people who consistently battle a weakness, you might feel alone when discussing it with others. The particularly tidy have a hard time relating to the messy. Extremely talented financial analysts have a hard time connecting with artists. We all have weak spots. You have aspects in your life you cannot fully overcome.

And that is just fine.

Have you ever wondered why some companies that start with two founders are much more successful than those that start with just one? Have you ever wondered why some of the greatest artists in history—such as The Beatles or Queen—have more than one founding member? While one person may excel at something, the others can fill in the gaps. It takes a community to support each other.

The weaknesses you have aren't the end of the world. But, if you don't find someone to offset your weaknesses, it can be very difficult to fulfill your ambitions. So, start working with other people or outsourcing expertise from others.

If you truly hate cleaning, consider hiring somebody to do it for you, especially if you have the resources to do so. If you want a beautiful piece of art that you could never create yourself, commission an artist to make it for you. If you want to dive into the financial stock-trading scene but don't know anything about stocks, contact a stockbroker.

You're not creating less value in your life by giving someone else a stab at your weaknesses. You're simply acknowledging them and finding solutions. You can't be good at everything. There simply isn't enough time in the day.

Once you've freed up your time by hiring people to counteract your weaknesses or simply allowing others to help, consider exploring what else is out there. You might assume that you're not good at something just because you haven't experienced it yet, but what if it's simply a hidden talent?

Finding your weaknesses and strengths is worth the time. It's all part of knowing yourself, and that comes with its fair share of bumps and bruises.

Reinventing Chicken

Harland David Sanders was born in the latter part of the nineteenth century. His father passed away when Sanders was just six years old, and Sanders took over many of the responsibilities of his mother when she went to work. He gained a particular interest in cooking food.

Later in his life, he took on a variety of roles, including streetcar conductor, farmer, and even fireman for the railroad. But even with all of his work, he never seemed to find something he truly desired. It wasn't until he turned forty that he found his calling.[44]

Sanders started cooking at a Shell service station in Kentucky. His cooking was very popular, which led to the opening of a small restaurant across the street. The growing success was tainted by severe setbacks, including the Great Depression and a devastating fire that destroyed his restaurant. Still, he believed in himself, and he rebuilt his restaurant. He was determined to refine his chicken recipe.

Finally, at the age of sixty-two, Sanders decided to focus entirely on his fried chicken, which included developing a pressure frying method that cooked the chicken quickly without sacrificing taste. He added a secret blend of eleven herbs and spices, which became the hallmark of his fried chicken.[45] He began traveling across the United

---

44   "Colonel Harland Sanders," *Biography*, last updated April 24, 2020, *https://www. biography.com/business-leaders/colonel-harland-sanders*. Sanders worked as a farmer, railroad fireman, streetcar conductor, and insurance salesman before finding his calling.

45   *Britannica*, "Harland Sanders," last updated December 12, 2025, *https://www.britannica.com/money/Harland-Sanders*. Sanders used the pressure cooker to keep the chicken moist while maintaining the flavor.

States, franchising his chicken recipe to restaurant owners and earning a commission for each piece sold. It worked out in the end, and his chicken recipe initiated a new chain called Kentucky Fried Chicken (KFC). From then on, he changed his name to Colonel Sanders and marketed himself with a white suit and black string tie.

You're not alone if you find yourself drowning in possibilities. You may think you know what your true passions are, but what if you're mistaken? What if you spend your entire life devoted to skills that won't serve you correctly? And what if your fear is what's stopping you from finding your true purpose?

It's never too early or too late to experiment. Many people attribute the search for self-discovery and trying new things to the young, but maybe you should push that aside.

If there's something that you've always wanted to try, especially if it's associated with providing an income, try it. Sometimes, the things that scare you the most are the most important for your personal journey. So, if you find yourself questioning whether you should take that job, try that new experiment, or set out on that new trail, just do it. You might find yourself at the end of it.

Chapter 9 Questions

- *What are some of the weaknesses you have not identified as such and just accepted because they have always been part of your activities?*

- *What strengths did you discover about yourself, and what plans do you have to expand on them?*

- *What value can you bring to other people by just being yourself?*

CHAPTER 10

Experience What You Desire to Experience

Knowing yourself is only half the battle. Now, it's time to put those principles into practice. Instead of telling yourself that you'll get to it "one day," it's time to start making what you want a priority.

Of course, what you want doesn't have to be completely separate from how you earn money. In fact, the best way to live an enjoyable life is to tie your desired experiences to your expertise. You're beginning to move in a direction more conducive to a happy life.

One of the biggest mistakes we make in pursuing our dreams is to assume that we need it here, that we need it now. Most of the time, the experiences you desire take time. You're not going to get everything you need to grow in your twenties. In reality, most people don't start making significant progress in their professions until they're in their thirties or forties. And there's no shame in taking your time to make the most of your experiences.

Nailing down your desires early helps you experience the right things to push you in the right direction. The ultimate goal of a happy life is to experience what you desire to experience, so always guide your decisions based on what will truly make you happy.

For example, if you've always had a desire to ski, why not try to make that something you can consistently accomplish? Live in a place that grants you more opportunities to ski. Find work that, even if it doesn't directly relate to skiing, will give you the time to do what you want. Or do a feasibility study to develop a business, related to skiing or not, that will allow you to live near to where you can ski. Even though you're constantly building the future that you want, nothing should stop you from enjoying where you are now.

In the meantime, grow. Gain financial resources so you can start doing exactly what you want to do.

Pursuing an Old Dream

By the time you reach the middle of your senior years, it becomes increasingly more difficult to maintain the lifestyle you could when you were younger. So, as you're exiting your seventies and entering your early eighties, you start to better understand your limitations. Your drive isn't what it used to be. Most of the time, it's tempting to put aside all dreams in favor of a quieter life. But, just because you feel a limitation, that doesn't mean you can't live life to the fullest of your ability.

Glance forward in years.

You sit on a rocking chair outside your retirement home. You want to make it to the store before it closes. You spent so much of your life wishing you had more time to fish, and now that you do, you've developed other limitations. So, you sit and wait for your son to come pick you up to take you to the tackle shop.

Your mind starts to wander as you sit and rock in that familiar chair. It's quiet here. The only sound is the muffled honk of geese in the pond around the corner.

Though it takes some effort, you hoist yourself out of the chair and walk inside. You pick up that old copy of Marcus Aurelius's *Meditations* and open it to the page that's been dog-eared for years now. Your finger traces the familiar quote, "You have power over your mind—not outside events. Realize this, and you will find strength."

At first glance, the quote had always seemed simple. But, as you get older, you've started to realize the power behind that kind of control. If you focus hard enough, you can completely detach from your external circumstances. Those are beyond your control anyway. You've seen your own limitations. You can't get around as easily as you used to, and it's more than a little frustrating that you have to wait for somebody else to help you.

But it's time to let all of that go.

Your son finally shows up, hours later than he had initially promised.

"Sorry," he says. "I got stuck late at the office. They're changing management, and I got stuck in the middle of it. You know how it is."

You heave a sigh, but then you smile. Yes, you do know how that is. But it's a life you gave up long ago. And right now, even though it's getting late and you may have missed the store's opening hours, it doesn't really matter. The hours you spent in wait weren't wasted. It's taken you a long time, but you've transitioned yourself from a kingdom of fate to a kingdom of freedom.

The tackle shop can wait. Right now, you're far more concerned with making the most out of every minute.

"Sit down for a minute. Let's talk about how things are going. And maybe I can shed some light on what it was like raising you."

Though we often associate limitations with growing older, we begin to fill these limitations throughout life. At times, we are unable to take exotic trips because of our limited income. At other times, we're limited simply by time. Rushing to children's events and engaging with friendships puts a significant limitation on personal time.

These experiences can be frustrating, realizing that there's only so much that we can do.

But true happiness starts to grow when we realize that it's all part of the journey. No matter where we are in life, we can always get closer to achieving what we want. Limitations will happen at any point in life, so it's best to enjoy the journey wherever you are.

The Heart of an Explorer

Marco Polo grew up in a family of merchants who engaged in trade in the Mediterranean and beyond. To him, receiving goods from the unknown world was fascinating, and it made him wonder how much was out there. So, when his father and uncle planned an expedition to Asia, he eagerly joined.

The Polos made their way east along the Silk Road, a web of trade routes connecting Europe to Asia. Their goal was to meet with Asian leaders and, hopefully, send messages from Pope Gregory X to Kublai Khan, the leader of the Mongol Empire. Along the way, they would explore new cultures, traditions, and places in Central Asia that no one from Europe had seen before.

When Marco Polo returned to Venice in 1295, the experience had so changed him that he insisted on sharing what he had learned. He wrote down a trove of stories and observations from his years abroad, creating a book called *The Travels of Marco Polo*.[46] It gave Europeans a glimpse into the East, sparking curiosity in whoever read his words.

The fascination with sailing to the Indies instead of traveling directly through warring nations influenced several other explorers. In 1492, Christopher Columbus believed he could reach Asia by sailing west across the Atlantic. Just five years later, in 1497, Vasco da Gama sailed east around Africa rather than west. His successful voyage opened India up to a direct sea route to the East.[47]

Just a few years later, another explorer who had always been fascinated by Marco Polo's stories set sail in search of new paths to Asia. Ferdinand Magellan had been inspired by those who came before him to try something completely new. He knew quite well that it was impossible to sail between North America and South America, but maybe there was a way around them.

By the time he set out, Magellan already knew about the South Atlantic. It was well documented that the waters on the southern end of South America had strong currents and unpredictable weather, often resulting in extreme storms. Sailing that far south meant increased risks of running into ice and strong winds, which could tear ships apart. But he suspected there might be a way to navigate the islands at the southernmost tip of South America.

Though the journey resulted in Magellan's five ships dwindling to only one, and Magellan himself didn't survive the journey, he created his own legacy with that one trip, significantly increasing the size of

---

46   Marco Polo, *The Travels of Marco Polo* (André Deutsch, 1959). Marco Polo described his experiences from 1271 to 1295. The book was written as a whole by Rustichello da Pisa from the accounts of Marco Polo.

47   *Britannica*, "Vasco da Gama," by Eila M. J. Campbell and Felipe Fernandez-Armesto, last updated December 20, 2025, *https://www.britannica.com/biography/Vasco-da-Gama*. The voyage opened the sea route to more Europeans by navigating around Africa.

the known globe.[48] Countries from around the world were inspired, which led to an explosion of exploration.

Curiosity is a major human driver. Most of the time, we gain more curiosity when we hear the stories of others going on adventures and exploring the unknown. Had it not been for Marco Polo, it's possible that many people wouldn't have been so excited to try something new, to go places no one had been before. He made so many people want to experience something based on his contributions.

Marco Polo's experiences, years before whole continents would be added to the map, changed the world. It's proof that anything that anyone experiences can have a vital impact on the world.

Determining Where You Want to Be

At the young age of seventeen, Marco Polo knew what he wanted out of life. Moving through exotic lands and experiencing new cultures was the most incredible lifestyle he could imagine. And his determination started when he had almost no means of achieving what he wanted.

It's easy to become discouraged when pursuing what you want in life because you are not at the level you want to be. Becoming a great baseball player comes with a lot of failed pitches and injuries. Becoming a scientist also comes with mostly failures. But everyone has to start somewhere.

While you're gaining experience, be satisfied with where you are by keeping an eye on where you want to be. Look for experiences that will come your way. The more you experience, the more you will start to understand where you want to be. Being an amateur is the perfect

---

48   *Britannica*, "Ferdinand Magellan," by Francisco Contente Domingues and Mairin Mitchell, last updated November 27, 2025, *https://www.britannica.com/biography/ Ferdinand-Magellan.* The trip was so harrowing that the entire mission remains shrouded in mystery, as some ships simply disappeared.

starting point for trying new things and experimenting. You'll soon find out what works and what doesn't, and you'll have fun learning along the way.

Creating Correlations

A lot of the pain that we suffer in life comes because of the lack of correlation we have between our individuality and the world. Often, that means that we don't stop to analyze our limitations before heading toward a goal. This is something many young people struggle with, but they're not the only ones who have mismatched their desires with their efforts because of the low level of correlations they've established between themselves and the world.

Everyone has a passion for something. We have a strong desire to accomplish something. It's why so many people in history have explored the unknown or climbed mountains simply because they were obstacles to overcome. Passion for something is one of our basic emotions, and it's the motivation to get you where you want to go.

The first step to finding your true goal is to learn more about yourself. What makes you happy? What do you desire to experience? What do you pay most attention to? How important is it to reach that goal?

We start to answer these questions by analyzing the reason why we're attempting to reach those goals. The first question to ask yourself is, *What do I want to accomplish or experience for a happy life?* Analyze your answer by asking yourself why you want to reach that goal. The more you question your motivations, the closer you are to understanding what your true purpose is.

If, for example, it's your goal to make $1 million per year, but you don't want to spend too much time working or learning a trade, you haven't found your true purpose. That goal is an outcome, something not directly tied to your passion or purpose. But if your goal is to make $1 million per year so you can spend more time with your family and start your own homestead, you are getting much closer to your passion and purpose. That $1 million marker is just a goalpost on your way to achieving your purpose and passion.

You're happy when you correlate what you want to achieve with what you want to experience. So, you may start a YouTube or TikTok channel that highlights what you do on your way to creating your homestead and spending more time with your family. You could also write your own book about your experiences or learn how to farm your own land. At that point, you're applying real-world applications to meet your criteria for a happier life.

Discovering More of What You Desire

Once you get a taste of new experiences, it becomes easier to find your path. New experiences provide the foundation for our emotional connections. For example, you might not have known you wanted to become a professional skier before you started skiing. In short, you'll never know what you're missing unless you try something new.

Many people have wondered for years if they'll ever find their true passions. So many people view the accomplishments of others from the sidelines and wonder how they figured it out on their own. The answer isn't as complicated as many seem to think it is: Those people continued to try new things.

Imagine, for example, you've never been to the ocean. You've seen pictures, and it's something that mildly interests you, but you always find yourself too busy to make the trip. Years go by, and you still think about the ocean occasionally. By the time you reach your seventies, you decide it's finally time to visit the place you've only seen pictures of. When you make it there, you see a rare glimpse of whales breaching the ocean. You make your way to the dock, and you see sea lions sunbathing on the rocks close to where you are. It's a breathtaking sight, and something you'd never imagined.

When you make it back home, you start spending your entire day investigating the ocean. You plan trips and even learn how to scuba dive. You tell your friends all about it and ask them to come with you on your next vacation.

Those years you stayed away from the ocean deprived you of pursuing a passion. You've become so much happier since realizing how important those unique experiences are.

The more you experience now, no matter where you are in life, the greater the chance you'll have of finding something that makes you happy. Passions are developed by people who aren't afraid to try something new. Happy people are always looking for something exciting, something that truly makes them feel complete.

And a truly happy life leads to the pursuit of more passions.

Achieving What You Desire

Getting those experiences starts small. If you don't know what you're most passionate about, sit down and analyze what you think most about. Where you're putting most of your attention is a key indicator of where you'll be happiest.

Many of us still feel like we can't experience everything we want to because of physical or financial limitations. Luckily, that doesn't mean we need to stop pursuing our passions or even finding out what they are.

There are always resources to help us achieve our goals. Throughout history, people have developed passions simply because of what they read by other people. Nowadays, you have many more ways to learn and explore, thanks to technologies already developed. So many people today, and people throughout history, have had the same goals as you. They each wanted to explore and grow. Resources do help, but such a desire to grow is not necessarily limited by the resources available.

For instance, Percy Fawcett became fascinated with the unknown and adventure at a young age. He read as many books as possible to fuel his imagination, hoping that someday, he would have the chance to explore. He became a British officer to travel the world and learn from new cultures. He learned how to survive in the wilderness and to navigate using maps and the stars.

While reading Sir Arthur Conan Doyle's *The Lost World*, he became convinced that he needed to see the Amazon rainforest for himself and look for lost cities. So, using his Royal Geographical Society connections, he visited the Amazon several times before eventually disappearing into the forest in 1925.[49]

His documented travels have fascinated people for a century. One reason he remains a legend is the documentation of his journey. He drew many maps and wrote down his experiences in a journal. All the while, he continued to have a preoccupation with the lost city of gold.

Many people have followed his example. Even if they never made it to the Amazon rainforest, they wrote many books about Percy Fawcett's experiences and their own speculations about what happened to him and the lost city. Before his last journey, he inspired several nations with the spirit of exploration simply because he had a passion.

Now, imagine how much further you could make it if you documented your own journey. You would see how far you've come and likely get inspiration to continue growing. Each new experience can light a fire toward a new passion.

---

49   Evan Andrews, "The Enduring Mystery Behind Percy Fawcett's Disappearance," *HISTORY*, May 29, 2015. *https://www.history.com/articles/explorer-percy-fawcett-disappears-in-the-amazon*. His son, joining his final expedition, disappeared as well.

Inspiring Others with Your Passion

A truly fulfilled life includes others. When you document your success, you're paving the way for others to experience your passions with you. And, even if they never meet you in person, if you document your journey, you can inspire countless others.

Leaving a legacy of where you've been and where you plan to go sets the stage for others to develop their own passions. Something that seems inconsequential to you might light the fire for somebody else.

We often don't see just how much we influence others. Whether it's a grand gesture or the simple motivation to keep trying to experience new things, you are probably an inspiration for someone. Anyone can seek their passions at any point in their lives. It often takes courage to get those experiences. So, it's logical to assume that someone wants to do what you have done, regardless of how little it may seem.

Accepting Art Through Limitations

Experiencing limitations is hardly the end of life. For Anna Mary Robertson Moses, also known as Grandma Moses, the discovery of her own limitations provided the spark for growth.

Grandma Moses had started life on the farm, and that's where she spent most of her life. She was a farm wife and mother with very little formal education, spending most of her time tending to her family and farm. In her spare time, she worked on embroidery, something she was known for locally.

In her late seventies, Grandma Moses developed arthritis, which made embroidering painful. It came as a crushing blow, especially since she spent so much of her time looking for ways to express herself creatively. But at some point, she had to acknowledge that she was no longer able to produce the intricate work she had dedicated her time to for years.

Finally, her sister approached her about painting instead. It was a creative outlet, which provided an escape. At first, Grandma Moses struggled. Arthritis had crippled her hand, making holding a paintbrush difficult, but she soon found ways to work around this problem.

Soon, she began depicting the life she had known growing up. She created folk art scenes of farm life, developing a unique style that included bright colors and simplistic locations. She used materials that were easy to come by and relatively cheap. She realized, in the late stages of her life, that sharing her creativity and memories made her happy. And that happiness came through in her paintings.

She decided to share her paintings at a local women's exchange at the age of seventy-eight.[50] And, though she hadn't set out to be famous, her charming style caught the attention of an art collector who introduced her work to the art world. By the time Grandma Moses reached her eighties and nineties, she had achieved international fame. Despite the simplistic nature of her artwork, she became an icon in American culture, and she continued to paint until her death at age 101.

Life is not perfect for anyone. Life is full of ups and downs, twists and turns. But nothing should stop you from trying to have more experiences. You'll never know just how much you can grow if you never give something new a chance.

Expanding our horizons is at the heart of pursuing happiness. The more we look outside ourselves and into something new, the more fulfilled we will feel. And that period of personal growth never truly ends. No matter the age, there is always something we can do to experience what we truly desire to experience.

---

50 *Britannica*, "Grandma Moses," last updated December 9, 2025, *https://www.britannica.com/biography/Grandma-Moses*. Grandma Moses loved the simple life, which she depicted in her drawings.

Chapter 10 Questions

- *What do you desire to experience? If you have more than one desire, you may list them in order of priority.*

- *What are the limitations you've given yourself because you're worried you can't fulfill what you want to?*

- *How are you putting your desires on hold, and how can you work with your limitations to experience what you desire to experience?*

CHAPTER 11
Push the Boundary to Serve the Purpose

Achievement is at the heart of who we are. It's why a simple task, such as making the bed, gives us a little boost of satisfaction. The more we try to accomplish, the happier we are.

Of course, that doesn't apply to everything. We can be happy when we get the chance just to relax and do nothing. We're not all hardwired to constantly improve our mathematical skills, draw, or scale mountains. However, when we want to achieve something, the reasons why we strive to accomplish that thing really matter. Once we accept their purpose and use it as a launching board, it's possible to accomplish virtually anything.

But, as with everything, there's a limit to how much we can accomplish. There are still mathematical problems that have yet to be solved. No one has yet mapped the Amazon rainforest. Humans have never left our solar system. And, in simple day-to-day life, sometimes there's just not enough time in the day to accomplish everything we want to. There are times when it's impossible to achieve what we want to because we don't have the right tools, knowledge, or natural abilities.

No matter how long you work at your passion, you'll always hit a roadblock.

The lack of progress can get frustrating. So many people have followed their purpose and passions as far as they were able, getting very close to completing goals but never quite reaching them.

Luckily, achieving all of your goals isn't necessary to be happy.

Many people, as they reach the twilight years of their lives, start to come to this conclusion. Albert Einstein spent years developing his theories of relativity. Isaac Newton continued to develop mathematics, dipping into alchemy in his later years. Leonardo da Vinci developed anatomical drawings and pioneered engineering projects for canal designs. All three of these men, though they had devoted their lives to science, found God in their final years. Letting things go when you've reached your limitation, putting them in the hands of a higher power, can be a common theme.

So, knowing you'll have limitations and that you'll never meet every boundary you strive to head toward, take a moment to relax. In the end, your goal is to get as close as you can to achieving as much as you can without crossing that boundary. After all, relaxing and doing nothing but just taking a breath to have a break can be enjoyable too, especially after you've tried so hard.

Happiness Despite Limitations

Reaching a point in life when your body starts to fail and your mind starts to drift more often is rare. People who make it past the age of eighty-five usually start to think of their futures in terms of tomorrow. It's impossible for them to know just how much time they have left.

Take a final trip forward in time, looking at the final days of your senior years.

You glance over at the crackling sparklers in front of you. They're blurry now. Your eyesight isn't what it used to be. In fact, nothing is.

You feel your daughter shake your shoulders, and you put a smile on your face as you glance around to see everyone smiling at you, singing the words to "Happy Birthday." Your daughter tells you that you're ninety years old. You smile, but on the inside, you're shocked. It's been a long time since you thought about your age. But it must be true. You're surrounded by so many people, and your granddaughter is holding her newborn son in her arms.

Despite the difficulty you feel when standing, you're grateful. Can it really have been that long?

"Tell us a story!" It's one of your grandsons. You've always liked him, and now, he's pursuing his own goals, traveling across oceans to climb mountains. It's something you did yourself all those years ago.

It's hard to remember all the details. You've told those stories countless times, but you can't seem to remember them the way you used to. Occasionally, someone will speak up, letting you know you missed a key detail. You correct yourself with an apology. Your mind just isn't what it was.

After the festivities, you breathe a sigh of relief. It's incredibly rewarding to see everyone again. But, at your age, it takes a lot out of you. Right now, even though it's only six in the evening, you just want to wind down.

Your daughter takes you to your bedroom and sets you down in bed. "Um," you mumble. "My memoirs. I was just working on them last night, and I know I don't have much time left. Is there any chance you remember where I put them?"

She smiles at you, though she looks sad. "Of course. I put them on your nightstand last night after you dozed off." She hands them to you and opens them up where you left off. "How many is that now?"

"I think I'm up to nine books now," you say.

"That's incredible."

"I don't know that I've lived an incredible life. There were definitely times I made the wrong decisions and did the wrong things. But, maybe if one of the grandkids or great-grandkids can get something out of this, it'll be worth it." You smile up at her. She's been so kind to you. She's been such a ray of sunshine her whole life. But now, her hair has gone gray, and it can't be easy for her to get around anymore. You know you won't be around much longer. You're expecting to go anytime now. But it's been a life worth living. And it's all because of the people in it.

Whatever stage you're at in life, no matter the age, life needs meaning. Problems arise when that purpose starts to fade. Without a purpose, life can become difficult to handle. And you're not guaranteed purpose throughout your life.

But there's always hope. Just because you won't be handed a purpose throughout life doesn't mean you can't create it on your own. Documenting your journey puts your experiences in perspective. You may start to see a pathway form, one where you're guided by internal desires or external pressures.

It's hard to see these connections in the beginning. That's why the documentation of your life is so important. Even though you might not connect the dots to very valuable lessons learned along the way, your posterity might. If nothing else, people will likely see your struggle to maintain purpose on the way to finding their own.

In the end, no matter how you live life, your story can spark purpose in someone else's life, even if your actions seem relatively small.

Sitting for Something

On December 1, 1955, Rosa Parks did something she thought was small.

She had just finished a shift as a seamstress in the local department store when she boarded the same bus she had for years. Parks made her way to the center of the bus, sitting in the first row of the "colored section" in the middle of the bus. Others followed suit, filling in the seats next to her. She knew about the segregation laws, and the injustice of them prompted her to sit up a bit straighter as she waited for White passengers to board.

Inevitably, the bus filled up quickly. When it was clear that there were no more seats for White passengers, the bus driver made his way to the middle of the bus. The bus driver asked for the first row of the colored section to move to make room. The other passengers in the front row quietly stood up and made their way to the back of the bus. Rosa Parks, on the other hand, simply sat and refused.

Frustrated, the driver demanded, "Why don't you stand up?"

Rosa Parks responded with, "I don't think I should have to stand up."[51]

After threatening to call the police, the driver still hadn't convinced her to give up her seat, so he followed through on his threat. Minutes later, Rosa Parks was arrested for violating Montgomery's segregation laws.

Rosa Parks had been a member of the NAACP since 1943. Years before this simple act of defiance, she had been determined to fight racial injustice. It was a small act of defiance, but it was deliberate and calculated from years of frustration with racial inequality. But she had no idea just how far the story of her defiance would spread. A single

---

51   "The Woman Who Changed the World by Sitting on a Bus," *Human Gene, https://www.humangene.org/en/article/the-woman-who-changed-the-world-by-sitting-on-a-bus.* Rosa Parks wasn't tired, but she was tired of giving in.

act inspired millions and prompted people around the country to discuss the implications.

The remaining years of her life were filled with contention. She was forced to flee her home in Montgomery for Detroit in 1957 because of death threats and employment discrimination. But her persistence on issues of racial injustice blossomed into other issues involving housing discrimination, police brutality, and supporting political prisoners. She developed a legacy in the form of mentoring young activists in her Rosa and Raymond Parks Institute for Self Development program labeled Pathways to Freedom. She remained dedicated to the fight until her death at the age of ninety-two.[52]

With one single act, she inspired generations.

Years later, Rosa Parks hadn't completed everything she wanted to in life. She had lived through the civil rights era and seen a significant increase in social justice awareness. Yet, there were battles to fight that extended beyond her capability. She was staunchly against the Vietnam War, but that part of her life usually pales in comparison to her influence when she was in Montgomery. She was a staunch opposer of apartheid in South Africa, a place where she had virtually no voice. Despite her popularity, she experienced her own limitations.

Understanding Limitations

Limitations can be hard to accept. We want to do as many things as we can before our time is up. It's frustrating to acknowledge that we might never see the end of our goals.

But there's a purpose to those limitations. Sometimes, we're guided toward a purpose in the form of those limitations.

---

52 "Pathways to Freedom," The Rosa and Raymond Parks Institute for Self Development, *https://www.rosaparks.org/programs/*.

Imagine, for example, you find a job that fits your background. You are excited about the wage, and the opportunities to grow make you more excited about this position than any you've seen before. Best yet, even though you have already started creating your own business, this job would finally give you a steady income stream. This opportunity is a lot less terrifying and a lot more stable than running your own business. After going through several rounds of interviews, you start to tell people that it is only a matter of time before you start work. You begin buying new clothes in anticipation of your new job's needs. You set your business idea aside, excited that you will start making a lot more money.

But, weeks later, you get an email you've gotten many times before: "Thank you so much for applying and interviewing with us. We spoke with many incredibly talented people, and unfortunately, you were not selected for the position."

The results seem devastating, and you're more frustrated than you were when you received similar emails. Your qualifications were a limiting factor. At times like this, it's easy to feel down and disheartened.

It's unsatisfactory to acknowledge that there will always be limitations. Everything living has a limited reach, and even though we may desire to do as many things as possible, there's only so much that we can do within a limited lifespan and with limited resources.

Time on a Budget

Because our time is limited, we have to budget it as well as possible to ensure we're staying on track. There are two ways of looking at this.

First, your life is broken down into a limited series of accomplishments. No matter how many of these you plan for, you can't fulfill them all. The only logical solution is to cut back on goals to make them more achievable. Once you've reached your time budget, you'll have to stop anyway.

The second is a more optimistic view: Plan as much as you can to meet your budget limit, and enjoy the journey along the way.

The natural course of life means that there are some passions you can't pursue. There simply isn't enough time. But, if you can find something you truly enjoy, it makes your journey much more enjoyable. The point of discovery is to enjoy what you're doing. So, if you find yourself overly stressed, stop what you're doing. There's no need to waste your budgeted time on something that won't bring you happiness.

Far too many people focus on passions outside the scope of their limitations. When it becomes increasingly hard to achieve those goals, stress naturally arises. It's like trying to fit a square peg into a circular hole. The more time you waste trying to fulfill passions that don't serve your purpose, the more stressed you'll become.

If you're too stressed, you halt your improvement. The limited time available to you isn't worth wasting on stress. Though it might be painful, stress may be a good indicator that you are not on the right path.

Meaningfully Pushing the Boundary

Imagine yourself in a giant bubble. The edges of the bubble represent your limitations and boundaries. As you look around you, you see a variety of passions that possibly lead to a fulfilling life. To your right, you see your passion for astrophotography, which encourages you to stay up late into the night and live in remote locations. To your left, you see a passion for urban foraging, combining your knowledge of plants with your desire for sustainability. Above you is your passion for historical reenactment, involving deep study of specific time periods and biographies. Below you is your passion for long-distance hiking, the desire to push your physical endurance as far as it will go.

These very distinct passions create a problem: They all require a lot of time and dedication.

You could try using each one of your limbs to push toward each of the different passions. But because you're extending yourself so far in each of the directions, you're not truly excelling at any of them. You're putting a physical strain on yourself by trying to push so hard in each direction. At some point, you have to stop pushing so hard to maintain your happiness.

Instead, if you use both of your hands to push as hard as you can in one or two directions, you'll start to see a shift in the bubble. You can move much more easily and even push beyond the boundary you currently have if you're putting your entire focus in one direction.

Luckily, that doesn't mean those other passions disappear completely. Pushing in one direction can get tiring, too. Pushing too hard leads to stress, no matter which passion you follow. So, when one limb gets tired, push in another direction. Just leave enough space for yourself to be happy exploring new passions.

Stopping to Rest

Moments of rest are vitally important for making the right decisions. A moment of silence provides the perfect atmosphere for evaluating where to go next. Sometimes, these moments take us in directions we never expected.

One man, Brandon Stanton, had a major change of heart during his period of rest. He had worked as a bond trader in Chicago for three years, becoming physically and mentally exhausted all the time. He worked long hours analyzing data and trading bonds, which often meant he slept at his desk. It became so unmanageable that he let his work slip, and he was fired from his trading job.

Even though he had spent much of his career in finance, the moment he stopped to rest and think about what he really wanted in life, he stopped pushing in the same direction. Using the last of his savings, he bought a camera and moved to New York City with the goal of photographing ten thousand New Yorkers.[53] He had no photography experience, but he needed to try something new.

In 2010, Brandon Stanton started a simple photography project he called Humans of New York. He uploaded the photos to social media and included short synopses of larger stories, often including quotes from unique people around the city.

He hadn't expected the rise of a global phenomenon.[54]

The stories touched people around the world, which prompted humanitarian fundraising campaigns for various causes. He wrote books about his experiences that reached global audiences. Today, he sells books using the portraits he collected over the past fifteen years to describe the soul of a city.

Making a radical change meant that Brandon Stanton found his true purpose. Recognizing his limitations after years of unhappiness was the perfect catalyst for making a switch.

That brief moment of reflection is imperative for helping you readjust. The more adaptable you are in adjusting your fundamental beliefs, the more likely you are to experience new things, ultimately pushing in new directions. And there's never a better time to do that than when you're tired. It leaves you more open to pursuing things

---

53   Brandon Stanton, *Humans of New York* (St. Martin's Press, 2013). Stanton says that he'd taken thousands of portraits and was broke before he got traction.

54   Brandon Stanton, *Humans* (St. Martin's Press, 2020). Stanton debuted at number one on *The New York Times* Best Sellers list when he presented a book with stories from over forty countries.

that you might not have considered before. It's a way to heal yourself guided by your passions and desires.

After you've made a switch, consider the change in your belief system. It's possible the path that led you toward stress was based on beliefs that were not developed under the right circumstances. If you're happy moving in your new direction, adjust your belief system and give yourself permission to adjust in the future.

Making adjustments from time to time makes it easier to be open to new beliefs and to recognize falsehoods, which is the perfect way to maintain a happy journey.

Running On

Madonna Buder had a rather traditional upbringing. She was born in 1930 to a Catholic family. After attending Catholic schools, Madonna took her vows, becoming one of the Sisters of the Good Shepherd at the age of twenty-three. For the first twenty-five years she spent in the convent, her life consisted of teaching, community service, and adhering to structured religious life.[55]

At the age of forty-eight, Sister Madonna spent a religious retreat with Father John Topel. He had been a runner for years and suggested that the activity worked perfectly with prayer and meditation. It perfectly blended body, mind, and spirit. It was also a chance to exercise both the body and mind. So, with no prior experience, she started jogging.

She started at a local high school track and branched out to other locations as the years went by. Over time, it became a passion. She

---

55   Sister Madonna Buder and Karin Evans, *The Grace to Race: The Wisdom and Inspiration of the 80-Year-Old World Champion Triathlete Known as the Iron Nun* (Simon & Schuster, 2010).

began participating in marathons in her fifties, and at the age of fifty-two, she completed her first triathlon. At age fifty-five, she completed her first Ironman competition, swimming for 2.4 miles, biking for 112 miles, and running for 26.2 miles.[56]

By the time she was in her eighties, Sister Madonna had competed in over 340 triathlons and 45 Iron Man competitions, leading to her appearance in a Nike advertisement at the age of eighty-six.

As she aged, she received criticism for her time in the spotlight. Her age, combined with her duties in the convent, seemed like an unusual match. But she had developed such a passion for physical activity and personal meditation that the criticisms simply didn't matter. She never made it about age. For her, it was all about attitude. Her two passions, religion and athleticism, worked together to help her create a more transformative life.

Despite her journey of injuries, setbacks, and the natural challenges of aging, she felt more fulfilled than ever.

We'll never truly know the boundary of our limitations if we never look for it. It's easy to become distracted by good or bad alternatives. That's why experiencing purpose is so important. It gives us guidance for the future.

Getting tired at some point is inevitable. Being burned out, even with the things that you love to do, causes stress and shows you that you might want to try something new. Despite the number of goals you hope to achieve in life, it's impossible to achieve every one of them, so find happiness along the way.

When you're tired, free yourself by letting go and pushing in a different direction.

---

56   "Iron Nun: Madonna Buder's Extraordinary Athletic Journey," *Inspiring Athletes*, July 6, 2024, *https://inspiringathletes.com/madonna-buder/*. She claimed to "train religiously," finding the perfect harmony in mind and body.

Chapter 11 Questions

- *What limitations have you held in your belief system that were created by yourself or someone else?*

- *When was the last time you were very stressed and stopped to ask yourself if you were headed in the right direction?*

- *What belief systems should you change based on inconsistent data?*

CONCLUSION

Born in 1838, John Muir showed a remarkable aptitude for mechanical creation. When he was a young man, he enrolled in the University of Wisconsin–Madison to study industrial advancement. He found success quickly by creating time-saving devices and working in factories to promote innovative machines to employers. He became completely absorbed in his work, often meeting with factory leaders to express his ideas.

Everything changed in 1867 when he had an accident that left him temporarily blinded.

Though he had spent most of his life working with machinery, he began to reevaluate his priorities. He began by walking from Indiana to the Gulf of Mexico.

The journey changed his perceptions of ambition. He found happiness in walking through nature. It was an abrupt shift from his background of industrialization, but he had finally found his purpose in nature.

Muir eventually made his way to California's Yosemite Valley. The landscapes were so beautiful, he believed that, at last, he had found his purpose. Rather than pursuing wealth or status, he would finally dedicate himself to exploring, documenting, and protecting America's wildernesses. He would spend his entire life in nature if he could. In a subsequent journal entry, he said, "I am losing precious days. I am degenerating into a machine for making money. I am learning nothing in this trivial world of men. I must break away and get out into the mountains to learn the news."[57]

His friends noticed a significant change in his advocacy for nature. He became far more fulfilled than others who had achieved material success. In fact, one acquaintance even described him as "a man who appeared to be happy all the time."[58]

Muir became known as America's most influential conservationist, befriending people such as Theodore Roosevelt.[59] He lived a life of contentment by aligning his values to his purpose and becoming truly happy.

The Heart of Correlationism

John Muir completely transformed his life through a framework I call Individual Correlationism, which is examining your life through correlation with your very individual experience, whether passive or interactive, direct or indirect, intentional or unintentional. Every moment, we're developing a relationship between what exists, what we know, and what we value.

---

57   John Muir, *My First Summer in the Sierra* (1911).

58   Charles E. Vroman, "Recollections of Early Days: John Muir at the University," *The Wisconsin Alumni Magazine* 16, no. 9 (1915): 557–62, *https://scholarlycommons. pacific.edu/muir-reminiscences/34/.*

59   Tony Perrottet, "John Muir's Yosemite," *Smithsonian Magazine*, July 2008, *https://www. smithsonianmag.com/history/john-muirs-yosemite-10737/.*

At first, Muir saw success in terms of industrial progress and material gain—values he had absorbed from the world around him. As he began to explore nature, he realized that his previous axiom was only applicable within certain contexts. The small choice to enter nature changed and expanded his life theory.

Every person listed in this book has illustrated Individual Correlationism in action. Each expanded their framework when faced with new experiences, refined their understanding through trial and error, and improved the correlation between their beliefs and reality.

And this framework works universally.

The Three Dimensions of Understanding

Individual Correlationism examines life through three interconnected dimensions from a philosophical approach traditionally categorized as follows:

Ontology, or what exists, is the reality of your world, your environment, and your circumstances. For Muir, this was the accident that made him temporarily blind, the journey he took from Indiana to the Gulf of Mexico, and the shift from the industrial world to the natural world. It was an expansion of what he had originally recognized as existing and maturing.

Epistemology, or what you know, encompasses your knowledge and understanding gained through experience. Muir learned to switch from understanding machines to understanding ecosystems, from knowing factory processes to knowing wilderness preservation, from knowing he was great with machinery to knowing he had trouble with machinery because of his blindness, and from knowing financial gain meant success to knowing he could enjoy life with wilderness preservation.

Axiology, or what you value, encompasses your value system, ethics, and priorities. Over time, Muir's values transformed from wealth accumulation to conservation, from industrial innovation to natural preservation, and from income and status to self-actualization.

But I cannot let you forget that I learned this from physics. In geometric optics, there's an object in *object space*, a lens, and an image in *image space*. The object is the reality that exists (ontology), the lens represents you (the observer), and the image represents your understanding (epistemology). It's the lens that forms the image based on the object, or you who determines what you get from the reality. Your judgment of the image's quality, whether it serves your goals and aligns with your purpose or not—or colloquially, "good" or "bad"—is your value system (axiology). Individual Correlationism takes a deep dive into the correlation between all of these elements and your individual perspective. The higher the correlation between your understanding and reality—or colloquially, the "smarter" you are—the more effectively you can navigate your life.

The Individual Perspective

Humanity is developed by building on knowledge over generations with no clear limit to our exploration, but each of us faces the constraint of an individual lifespan. So, developing your personal understanding of life becomes much more important for yourself. You can't wait for collective humanity to solve your problems or answer your questions. Furthermore, even as the development achieved by the collective humankind to date becomes available at your fingertips (thanks to artificial intelligence), you will have to learn it and internalize it through your very own eyes (of course, figuratively here). Therefore, the only way to live a successful life is to develop your own framework in your own lifetime, even though you may be "standing on the shoulders of giants."

And this is where we get into the weeds. Individual Correlationism, as a philosophical framework, differs from pure science. Science begins with axioms, or statements accepted as universally applicable. Individual Correlationism allows you to start with your beliefs, or *your* axioms, as mentioned earlier in this book—in other words, your personal understandings that serve as your working hypothesis for how your life works. In the long run, these beliefs aren't necessarily universal, but they guide your decisions and actions. They might even be universal for you without you knowing it consciously, as you take them for granted. Like science, your beliefs are tested through experience. When you encounter contradictions or limitations, your beliefs are altered to become more applicable to different conditions. And, in essence, you expand your framework to broader contexts.

As you navigate through life, your framework develops to encompass three progressive levels of engagement that apply to anything in life from an epistemological perspective:

I See: At its most basic level, you observe the world around you. You see a chair, a person walking, a situation unfolding. In Muir's case, he became actively engaged in his observations of nature on his walks. Observation is where all understanding begins.

I Know: Over time, your observations develop to encompass your experiences of life. You know whether the chair is for sitting on or for decoration, you know the person is an architect, and you know the situation requires action. Muir knew, through his industrial experience, that work provided income and status, which served as the extent of his framework before his transformation.

I Understand: With enough experience, you start to have a deep comprehension of how and why something works. You understand how the chair was constructed, why it has four legs for balance, and even how you could build one yourself. You understand not just what something is or what it does but also its underlying principles. Muir reached this level when he understood not just that nature existed and was beautiful but also why preservation mattered, how ecosystems functioned, and how to advocate effectively for conservation.

All three levels are universal. They can apply to objects, relationships, ideas, systems, or life experiences. The deeper you go into each level, the greater your agency and more effective your actions. The deeper your understanding, the more you can achieve, the better you can align your life with your values, or colloquially, the happier you can be.

How Your Framework Expands

Everyone develops their understanding through a process that mirrors scientific methodology:

1. **Experience:** You encounter new situations, test approaches, observe outcomes. These are your personal experiments.

2. **Theory formation:** You develop beliefs about how things work based on your experiences and personal theories about life.

3. **Application:** You apply these theories within their scope, using previous patterns you've recognized as blueprints.

4. **Expansion:** You begin to recognize the limitations of your theories when you encounter situations that expand your framework. Old beliefs become conditions that are still valid in certain contexts but aren't universal principles.

The cycle continues throughout your life, expanding all the time. You will always find inconsistencies as your scope expands. But your goal is never perfection. A life well lived has continuous improvement in the correlation between your understanding and your reality.

How to Start Your Practice

My philosophy begins with seeking greater clarity in one area, such as your career, relationship, health goal, or any other situation that matters to you.

First, observe without judgment. Find out what is actually happening. Gather experiences. Notice patterns. Don't rush to conclusions. Just learn how to collect data about your reality.

Second, develop your working theory. Learn to recognize patterns that emerge. What do you believe is happening, and why? No theory needs to be perfect, just testable.

Third, test and apply. Act on your theory. Can you achieve what matters to you, and does your prediction match the outcome? If the answer is yes, you've developed a strong correlation with reality in that area. If the answer is no, you've discovered valuable information about your theory's limitations.

Fourth, take note of the boundaries. Your theory won't always work. So, start to ask yourself when it fails and what conditions make it applicable.

Finally, expand your framework. Your original theory serves as the basis for broader understanding. You've elevated your knowledge to include more contexts, and it's time to search for more.

This process never ends. Every time you repeat this cycle, you're improving your correlation with reality. Every time you expand your framework, you receive more tools for navigating life effectively.

The Power of Correlation

Correlations are never perfect. There will always be some limitations, some aspects of life beyond your control or understanding. But there are always ways to improve the levels of correlation over time. The secret source of your achievement and happiness is the knowledge to push the boundary when you can and rest when you need to.

The key to understanding this framework is the correlation between your theories and reality. Higher correlation means your theories effectively work. The more effectively your theories work, the higher your achievement and happiness.

But where do you draw the line when it comes to what's in your control and what isn't? Understanding the correlation between your theories and your actual situation gives you the best understanding of what you control and what you don't. The higher the correlation, the clearer you can be about what's in your control and what's not.

But perhaps one of the most important principles to remember is that there are unlimited perspectives on any situation. Your understanding, no matter how refined, will always be incomplete. And that's the perfect invitation to remain curious. Others see what you can't. Their perspectives aren't necessarily wrong, merely different from yours. The more we grow our perspectives by collaborating with others, the greater our happiness.

No one exists in isolation. We all live in a society where we interact with others who are also seeking to understand their lives and achieve their goals. Your freedom to develop your understanding and pursue your happiness should never come at the cost of someone else's freedom to do so. Always remember that any other individual is exploring their life for their achievements and their happiness, just like you are.

Understanding this naturally gives rise to benevolence. Seeing others engaged in the same process of understanding their lives by testing their beliefs, refining their frameworks, and seeking better correlation with their reality, it's easy to start extending the same patience to them that you give yourself. Benevolence isn't just kindness but the recognition of our shared human condition of imperfection.

We grow as individuals, but we excel at communities. Your story, however ordinary it might seem to you, could influence others in ways you would never realize. When you combine your personal growth with collaboration with others, you can achieve far more than you ever could in isolation.

The Journey of Understanding

Fulfillment comes at any point in life. Your mistakes don't define your future at any age. It all depends on developing healthy perspectives that help you to continue to grow. It's why happiness is best defined as satisfaction when you eventually get what you wanted. Once you match your experience with what you desire and dedicate your life of purpose to fulfilling your desires, life becomes much simpler and much happier.

Socrates said that "an unexamined life is not worth living." This is the part of his perspective on life—or his value system, if you will—that I very much share. Individual Correlationism provides the methodology to test your beliefs, expand your understanding, and improve the correlation between your internal frameworks and external reality.

Life is short. So, the faster you learn to develop your personal framework, the better you can understand reality, and the more you can achieve and experience in your lifetime.

No matter where you are in life, it's easy to believe that success depends on material or external validation. But fulfillment often comes from a different source—from aligning what you understand with what exists, from matching what you value with how you live, and from continuously expanding your framework as you grow. This is not to say that this framework of Individual Correlationism won't work if you continue to believe in success based on material or external validation.

So, here is my offering: a framework you can adapt to your individual life. Find your purpose by examining your experiences. Develop your theories by recognizing patterns. Test those theories through new experiences. Expand your framework when you see limitations. Improve the correlation between your understanding and reality.

And throughout this entire journey—

Enjoy it. Have fun!

The Joy of Living

I see, I know, I understand,
Each truth unfolding in my hand.
I try, I stumble, yet I rise,
With clearer heart and wiser eyes.
For each attempt reveals some more,
A deeper knowledge than before.
Again I strive, again I bend,
Each failure but a patient friend.
Perhaps success will come to me,
Or failure stay—my old decree.
But joy abides in all I do,
The journey is its triumph, too.